T0201318

Smart Water Technologies and Techniques

Challenges in Water Management Series

Editor:

Justin Taberham
Independent Consultant and Environmental Advisor, London, UK

Titles in the series:

Smart Water Technologies and Techniques: Data Capture and Analysis for Sustainable
Water Management
David A. Lloyd Owen
2018
ISBN: 978-1-119-07864-7

Handbook of Knowledge Management for Sustainable Water Systems
Meir Russ
2018
ISBN: 978-1-119-27163-5

Industrial Water Resource Management: Challenges and Opportunities for
Corporate Water Stewardship
Pradip K. Sengupta
2017
ISBN: 978-1-119-27250-2

Water Resources: A New Water Architecture
Alexander Lane, Michael Norton and Sandra Ryan
2017
ISBN: 978-1-118-79390-9

Urban Water Security
Robert C. Brears
2016
ISBN:978-1-119-13172-4

Smart Water Technologies and Techniques

Data Capture and Analysis for Sustainable Water Management

David A. Lloyd Owen
Envisager Limited
Trewindsor Farm
Llangoedmor
Ceredigion
UK, SA43 2LN

Registered Offices
John Wiley & Sons, Inc., 111 River Street, Hoboken, NJ 07030, USA
John Wiley & Sons Ltd, The Atrium, Southern Gate, Chichester, West Sussex, PO19 8SQ, UK

Editorial Office
9600 Garsington Road, Oxford, OX4 2DQ, UK

For details of our global editorial offices, customer services, and more information about Wiley products visit us at www.wiley.com.

Wiley also publishes its books in a variety of electronic formats and by print-on-demand. Some content that appears in standard print versions of this book may not be available in other formats.

Library of Congress Cataloging-in-Publication Data has been Applied For
ISBN - 9781119078647

Cover Design: Wiley
Cover Images: (Foreground image) © Tetra Images/Gettyimages;
              (Background image) © ESB Professional/Shutterstock

Set in 10/12pt WarnockPro by SPi Global, Chennai, India
Printed in Singapore by C.O.S. Printers Pte Ltd

10 9 8 7 6 5 4 3 2 1

Contents

Series Editor Foreword – Challenges in Water Management

The World Bank in 2014 noted:

'Water is one of the most basic human needs. With impacts on agriculture, education, energy, health, gender equity, and livelihood, water management underlies the most basic development challenges. Water is under unprecedented pressures as growing populations and economies demand more of it. Practically every development challenge of the 21st century – food security, managing rapid urbanization, energy security, environmental protection, adapting to climate change – requires urgent attention to water resources management.

Yet already, groundwater is being depleted faster than it is being replenished and worsening water quality degrades the environment and adds to costs. The pressures on water resources are expected to worsen because of climate change. There is ample evidence that climate change will increase hydrologic variability, resulting in extreme weather events such as droughts floods, and major storms. It will continue to have a profound impact on economies, health, lives, and livelihoods. The poorest people will suffer most'.

It is clear there are numerous challenges in water management in the 21st Century. In the 20th Century, most elements of water management had their own distinct set of organisations, skill sets, preferred approaches and professionals. The overlying issue of industrial pollution of water resources was managed from a 'point source' perspective.

However, it has become accepted that water management has to be seen from a holistic viewpoint and managed in an integrated manner. Our current key challenges include:

- The impact of climate change on water management, its many facets and challenges – extreme weather, developing resilience, storm-water management, future development and risks to infrastructure
- Implementing river basin/watershed/catchment management in a way that is effective and deliverable
- Water management and food and energy security
- The policy, legislation and regulatory framework that is required to rise to these challenges
- Social aspects of water management – equitable use and allocation of water resources, the potential for 'water wars', stakeholder engagement, valuing water and the ecosystems that depend upon it

This series highlights cutting-edge material in the global water management sector from a practitioner as well as an academic viewpoint. The issues covered in this series are of critical interest to advanced level undergraduates and Masters Students as well as industry, investors and the media.

Justin Taberham, CEnv
Series Editor
www.justintaberham.com

Introduction

My involvement with smart water stems from a project examining smart water policy drivers for the OECD (Lloyd Owen, 2012a) and as part of a more general study on urban water services (Lloyd Owen, 2012b). These considered the evolution of smart water as a concept and in reality especially between 2011 and 2012.

'Smart water' is not a theory, let alone a paradigm. Rather, it is a catch-all expression that covers real or near real-time data collection, transmission and interpretation for improving the delivery of water and wastewater services and optimising the performance of the assets that are used for these. This study is practical in nature, outlining what smart water means from various water management perspectives and how it has been developed and deployed to date. Much of the information is derived from conference presentations and articles in water sector publications rather than academic publications. This book is neither a technical nor an academic study. Instead, it considers smart water's potential to address a range of challenges currently facing water and wastewater management worldwide. Market drivers are reviewed along with the markets themselves, their size, growth and social, regulatory and environmental drivers. This book considers how the practicalities and prospects of smart water as perceived in 2016–17.

Making technologies work matters. Despite considerable regulatory, financial and political support, slower than anticipated development of practical energy storage has delayed the widespread adoption of electric cars by more than two decades. Smart water hardware development has not seen such technical setbacks. The challenge for commercialising innovation in smart water lies in raising funds and encouraging its adoption in an inherently conservative sector.

The most notable change since 2011–12 has been in the way smart technologies are being applied. The rapid and hitherto unexpected rise of smart phones for example has transformed the scope for mobile smart water monitoring and analytics. In developing economies, this may bring about truly disruptive changes. If a second edition of this book is published at some point in the future, the changes and their impact in those countries are likely to be appreciably greater than those that have been experienced to date.

Another change since 2011–12 is the gradual replacement of theories with realities. A difficult investment climate has meant that a significant number of intriguing innovations seen in 2011–12 have fallen by the wayside. Some of this can be seen as the natural consequence of attrition, yet there is always the fear that genuinely useful innovations can be lost during a particularly hard period for early-stage companies. A contrary point of view would be that products and services which can reach commercial viability under these circumstances may have the potential to offer a real and lasting benefit to utilities and their customers alike, having proved themselves in such a testing environment. It is also noticeable that despite many setbacks and the usual challenges in bridging the gap between blithe optimism and cooler realities, smart technologies and their applications are being more widely adopted.

Industrial water will only be covered in passing. Given that industrial clients are usually more open to innovation than municipal clients, as they are driven by the need to carry out processes in the most efficient manner, this may appear to be anomalous. This is in part due to the fact that industrial facilities are regarded as stand-alone entities, rather than being parts of networks, even when they are connected to municipal supplies. Their operations are relatively compact and most water and effluent assets operate above ground, making physical inspections more effective. As industry is driven by the need to be efficient in order to be competitive; smart applications that can improve the value generated by each unit of water consumed will be adopted where needed.

Smart water is evolving in an appreciably faster manner than is usually seen for the development and deployment of goods and services associated with drinking water provision and sewerage and sewage treatment. As a book, this is therefore a child of its time. It aims to present how various goods and services were being developed and deployed at the time of its writing in 2015–17, in the context of the author's experience with the concept since 2011.

An overview of the 'trajectory' of the deployment of smart water products is offered through examining third-party surveys (Chapters 1 and 7) as to their future extent, and at the end there is an attempt to suggest where the various initiatives that have been described could lead us, in terms of a truly integrated water and wastewater management system.

This book would not have been possible without the support, insights and information that a wide variety of people have given me.

Xavier Leflavie and Gerard Bonnis oversaw my project for the OECD in 2011–2012. Sophie Treemolet and Bill Kingdom managed a project on capital efficiency at the World Bank in 2016–17 which provided insights into the potential for smart water in developing economies.

Three organisations have been of particular value in organising conferences dedicated to smart water: The Chartered Institute of Water and Environmental Management (CIWEM), the UK's professional body for water engineers; SWAN (Smart Water Networks) Forum, a UK based organisation dedicated to developing smart water; and SMi, a conference company that has hosted a series of smart water events. Presentations at their events have been indispensable for developing the case studies. Mark Lane at Pinsent Masons also deserves thanks for the 'Wet Network' events he has organised over the past decade along with support from Arup. Oliver Grievson (Anglian Water) has been a great evangelist for smart water as has Global Water Intelligence's Christopher Gasson, who has combined this with his desire to improve the quality of information about what remains a poorly understood business. Thanks are also due to Bruce Moeller (Aquaspy), David Henderson (XPV Capital), Professor Asit Biswas (Lee Kuan Yew School of Public Policy, NUS Singapore), Rob Wylie (WHEB), Jim Winpenny (Wynchwood), Michael Chuter (Pump Aid), Jack Jones (Sanivation), Philippe Rohner, Arnaud Bisschop, Simon Gottelier and March-Oliver Buffle (Pictet Asset Management),

James Hotchkies (JWH), Michael Deane (NAWC) and many others. Finally, Justin Taberham suggested that I write this book.

References

Lloyd Owen D A (2012a) Policies to support smart water systems. OECD, Paris, France.
Lloyd Owen D A (2012b) The Sound of Thirst: Why urban water services for all is essential, achievable and affordable. Parthian Books, Cardigan, UK.

1

What do we Mean by 'Smart Water?'

Introduction

This chapter considers and defines the terms and expressions associated with 'smart water' and places them in the context of water management in the broadest sense. It also presents a range of estimates and forecasts of smart water's market size and its share of the markets associated with water management and environmental goods and services in general.

1.1 Defining 'Smart'

1.1.1 'Smart' and Utilities and Public Services

When applied to utilities, environmental and public services, a working definition for 'smart' would be the application of data monitoring, transmission, management, and presentation to services in a manner that enhances the efficient use of their operating assets.

It covers data management and communications systems and services (ICT – information and communication technologies) for utilities public and environmental services. It can be seen as a catchier alternative to 'intelligent' which has also been applied here.

1.1.2 Smart Consumer Goods

In addition, 'smart' has been adopted for a wide variety of consumer goods. In November 2002, Microsoft announced that it was developing the Smart Personal Objects Technology (SPOT) Initiative, for 'improving the function of everyday objects through the injection of software' (Microsoft, 2002). While a range of devices were released by third party manufacturers (wristwatches, GPS navigation systems and weather stations) SPOT was discontinued in 2012, in particular due to the development of WiFi as a more efficient data transmission system (Gohring, 2008). Since then, 'smart' mobile phones, tablets, watches and cameras have been launched, along with TVs and cars under development.

Smart Water Technologies and Techniques: Data Capture and Analysis for Sustainable Water Management, First Edition. David A. Lloyd Owen.
© 2018 John Wiley & Sons Ltd. Published 2018 by John Wiley & Sons Ltd.

As will be discussed later, the migration of 'smart' into consumer goods such as washing machines, showers and lavatories is set to become a factor in domestic water demand management as the 'Internet of Things' (IoT) connects domestic devices into broader data networks.

1.2 'Smart Power' and 'Smart Grids'

Smart power management for electricity utilities has not been driven by one or a small series of dramatic or disruptive events; it stems from a gradual continuation of demand management approaches. Electricity metering for measuring electricity used was introduced in the 1880s and has been developed ever since, including the introduction of digital metering in the 1990s (Anderson and Fuloria, 2010). Smart electricity metering is being driven by utilities and legislation, especially in the European Union, where at least 80% of meters are meant to be smart by 2020 (European Union, 2009).

Smart electricity meters inform electricity consumers how much power they are using and how much this is costing. Differential daily tariffs can be exploited to take advantage of when it is cheaper to use electricity (the lower the peak level of demand, the cheaper it is overall to produce each unit of electricity) which in turn means that the utilities can smooth out their power generation more than when there is only a single tariff. This approach is a modern refinement of night storage heaters, which have been used for some decades, assisting users to consider when they use electricity for light, heat and hot water and to optimise the time when these are used, to smooth their power demand profiles.

1.2.1 Smart Grids

Electricity grids, whereby utilities link up various power generators into a network offering greater security of supplies and flexibility of capacity were developed in Europe and the USA in the first three decades of the 20th century. In the UK, the Electricity (Supply) Act of 1926 brought about the Central Electricity Board, which rationalised 600 local power generators into regional networks by 1933 which in turn were integrated into the National Grid in 1938 using the 132 most efficient power generators in the UK.

The smart grid is concerned with ensuring the most efficient use of electricity across a network, so that no more generating capacity is deployed at any one time than is needed, matching demand with supplies as closely as possible and ensuring that both the most appropriate generating capacity is deployed (using generators at their optimum output) and with minimal transmission losses. They are also intended to provide the most reliable service under given circumstances and more recently to lower the utility's environmental impact through renewable energy sources.

According to the Smart Grid Forum (Smart Grid Forum, 2014), a smart power grid is 'a modernised electricity grid that uses information and communications technology to monitor and actively control generation and demand in near real-time, which provides a more reliable and cost-effective system for transporting electricity from generators to homes, businesses and industry.'

Smart electricity grids were made possible by advances in data capture, communication and management through advances in computing, data transmission and metering in the 1980s and 1990s. The first major deployment was in Italy, where the Telegestore programme was launched in 1999 and was completed by 2006, resulting in a comprehensive smart grid and metering infrastructure. Efficiency gains meant that operating spending per customer fell from €80 in 2001 to €49 by 2008 (Drago, 2009). In social (Google) and technical (IEEE Xplore) media, the frequency of the use of 'smart grid' entries becomes increasingly frequent from 2008, with the first journal citation having taken place in 1997 (Gómez-Expósito, 2012).

1.3 Cleantech and Smart Cleantech

The expression 'Cleantech' ('CleanTech' and 'Clean Tech' are also used) is an abbreviation of clean technology. Cleantech covers goods and services that are designed to reduce the environmental impact of utility, environmental and public service activities such as power, waste management, heating and transportation, along with consumer goods associated with these such as washing machines and cars. Cleantech's driving principal is, wherever attainable, to 'do more for less' whereby an innovation both improves the performance of a utility or an allied service, and lowers its costs. It is thereby seen as helping to make essential goods and services more affordable while also improving the efficiency of goods and services and reducing wastage to a minimum.

In practical terms, Cleantech covers goods and services that maintain or improve productivity while lowering energy and material resource needs and lowering operating and manufacturing costs. This is typically brought about through improving efficiency, minimising the resource intensity and reducing the carbon footprint of these offerings. By bringing down the costs of these goods or services, their affordability is also improved, allowing for a more extensive adoption than was possible with traditional approaches.

In the author's experience, there have been three factors behind the term's popularity. Firstly, in the late 1980s, the expression 'environmental services sector' was initially adopted by the financial services sector for companies involved in waste management, environmental consultancy and contaminated land remediation. The water and sewage sectors were at the time regarded as utilities, and with some exceptions, the impact of environmental drivers on their activities had a low priority. Secondly, in the early 1990s, companies involved in providing environmental goods and services were considered 'recession resistant, if not recession proof'. During this period (for example there was a recessions in the USA in 1990–91, the UK in 1991–92 and Japan in 1991–93) it became evident that a decline in house building and decreased industrial activity did in fact significantly impact the environmental services sector and the expression lost its attraction to investors. Finally, the succinctness of the term and the way it allowed other applicable activities to be included (in particular, responses to climate change) made it an attractive expression for those involved with the industry in subsequent years.

Cleantech is in particular associated with aiming to decrease a product or service's environmental footprint, typically in terms of its CO_2 generation. The ultimate aim here is to 'de-carbonise' activities so that they are not net generators of CO_2. As a result, Cleantech is especially associated with developing and deploying renewable energy technologies.

1.3.1 Smart Cleantech

Smart Cleantech can be seen as an overlay of information processing upon extant systems. For example, the smart grid is the next stage of the adoption of smart Cleantech approaches, that of linking disparate activities together to that they can be monitored and managed in a more efficient manner than before. All aspects of Cleantech can potentially benefit from smart approaches where they enable the impacts of these innovations to be delivered in the most efficient manner.

Along with the smart grid, smart Cleantech is concerned with the automation of systems within Cleantech, managing their interfaces, ensuring that they are self-healing (for example, through negative feedback loops), by adopting integrated communications for monitoring, supervisory control and data acquisition (SCADA) and delivering usage optimisation and peak demand smoothing. These terms and their potential applicability will be considered in due course.

The principle of 'doing more for less' is particularly important in the water sector, which has greater funding challenges than other utilities. Smart water approaches will only be adopted if they allow water utilities and other users improved performance and service delivery and assist them to lower their capital and operating costs.

1.4 Smart Water

Smart water is a term derived from the 'smart metering' and 'smart grid' sides of the Cleantech industry for lowering electricity usage and making power distribution more effective and efficient. In terms of water, this covers water distribution and usage, wastewater distribution, treatment and recovery, and also covers water flows, quality and saturation in the built and natural environment. It is a concept that has been realised through the development and convergence of information technology, mobile and digital communication and the Internet.

Smart water 'is something of a catch-all expression' (OECD, 2012) for the current and potential impact of data collection, transmission and analysis for water and sewage utilities and domestic, commercial, industrial and irrigation users. As with the smart sectors previously described, smart water is in essence about achieving more while spending less. Despite being a part of water management in various forms for the past decade, in practical terms 'its definition and role remains a work in progress' (OECD, 2012). It is not intended to replace how services have been operated, rather to improve them and therefore to become 'an enabler of innovation, as much as being an innovation itself' (OECD, 2012).

As a concept, smart water emerged from 'Cleantech' in general and 'smart Cleantech' in particular, respectively as a suite of technologies designed to minimise and mitigate the impact of human activities on the natural environment and the potential for information technology, data transmission and perhaps, in the future, for using the 'Internet of Things' (IoT) to further optimise the effectiveness of such approaches. This is a somewhat radical approach for, as far as water management is concerned, it is a typically conservative activity and in consequence smart water is still at a tentative stage of its development. Indeed, its potential contribution towards addressing key structural challenges facing water and sewage management has not yet been fully appreciated.

A degree of caution is necessary, as it is often tempting to perceive an emergent technology or application as a realised one. Mobile communications provide a useful analogy. In the 1980s and early 1990s, mobile communications were seen as a dynamic and growing activity providing voice and limited data services at a high price to 10% of even 20% of the adult population in the more developed economies. Instead of being a premium service, mobile communications have since evolved into a low cost voice and an increasingly sophisticated data service whose coverage is becoming appreciably greater than that of fixed wire telephone services, especially in developing economies.

There are two ways of considering smart water. Firstly the parts of the water cycle that it can impact and how that impact may be felt and secondly, how it can influence the management of each of these components.

1.4.1 Smart Water and the Flow of Information

Smart water management typically involves five discrete stages in information handling. Data collection, interpretation and management may take place by using approaches such as JCS (data cache management for optimal data handling), CRM (customer relations management via dedicated data management), smartphones as data handlers and GIS (geographic information systems for collecting, analysing and sharing geographic information).

The examples of technologies involved below are in part based upon Heath (2015).

1.4.1.1 Monitoring and Data Collection

A monitoring system that enables the real-time (or as near to real-time as is practicable and needed) monitoring of all the necessary information for the effective management of the water service concerned and the collection of the relevant data. For example, in the water distribution mains this would include water flow and pressure, as well as temperature, pH, turbidity and the presence of treatment chemicals and contaminants. The monitoring data is then collected into a form that is suitable for its transmission.

1.4.1.2 Data Transmission and Recovery

The closer to real-time the data collection is, the greater the necessity that the data can be transmitted without human intervention. For example, the move from manual to automated domestic meter reading.

Getting data in from a number of remote sites covering a water or wastewater network, domestic customers or surface waters requires remote data transmission from the field monitors to the data management centre. This can be carried out through fixed wire or wireless data transmission. Mobile data approaches are driven by the cost of transmitting the data in relation to the value accrued from this information. High value data from a remote point justifies dedicated data transmission, while lower value data such as domestic metering can at the most basic level be gathered by, for example, a drive-by wireless data collection service.

Data communication may be 'piggy-backed' on to electricity or telecoms networks, through radio transmission, or various mobile data applications.

1.4.1.3 Data Interpretation

Data is collected at a monitoring centre and is processed so that it is in a useable form for its manipulation and presentation. Given the volume of data generated, this needs to be done on an automatic basis. One particular concern here is to ensure that all sources of potentially valid data can be accessed and that the system is open to accepting new data sources as they become available. The hybrid cloud (using private and public cloud-based data) may be used for integrating data from a wide variety of sources, such as water use, water demand, weather data and forecasts and monitoring external events what may affect water demand.

1.4.1.4 Data Manipulation
Data is interpreted according to each end-user's need. At this point, feedback loops may be used to feed new information into predictive models so as to be able to update any forecasts being generated and also to improve the model's predictive ability through the use of real-life information rather than simulated data.

1.4.1.5 Data Presentation
Finally, the information that has been gathered and analysed has to be presented in a manner which allows operators to act upon it in the simplest and most effective manner possible. This involves the use of graphics and alerts to inform an operator about any perturbations that ought to be of particular concern, while providing immediate access to the underlying data so that they can appreciate its particular nature. This may involve presenting information through a series of layers that allow operators to focus upon potentially relevant events and to locate and place them within their relevant operational context.

The first four stages can be seen as getting the data that a user needs, with the user acting on this data as presented in stage five. The object of stage five is to assist the user to make an informed decision based on this information. That may range from a domestic customer seeking to modify water usage to keep water (and electricity) bills down, a grower deciding when to irrigate crops or a utility manager considering which water resources to deploy.

The SWAN Forum (Smart Water Networks Forum, an industry group promoting the understanding and application of smart water management, swan-forum.com) defines data flow across smart networks (Peleg, 2015) as starting from the final outcome and working down to the infrastructure involved. They are as follows:

1) Automatic decisions and operations.
2) Data fusion and analysis.
3) Data management and display.
4) Collection and communication.
5) Sensing and control (including smart water meters).
6) Physical layer (including traditional and bulk meters).

Stages 2–5 are seen by the SWAN Forum as forming the smart water network.

1.4.1.6 From Top–Down to Bottom–Up; Inverting the Flow of Information
Smart water is redefining the way that information is gathered and in whose interest this information is gathered and where it goes. For example, data collection through smart apps on mobile phones allows people in developing economies to monitor their access to safe water and sanitation (and the presence or absence of open defecation) and send this information upwards, rather than relying on the traditional visitations of

government officials. Likewise, smart cash transfer approaches using mobile phones have both reduced customer time in paying utility bills and reduced the cost of billing for their utilities.

1.4.2 Smart Water and Managing the Water Cycle

Seven principal smart water applications can be identified. All of these are linked to some extent with the other elements.

1.4.2.1 Potable Water Systems

Optimising the beneficial use of water resources and managing water distribution networks to through minimising non-revenue water (NRW) and giving consumers tools to control their consumption, while maintaining the appropriate level of water quality and service delivery. This is delivered through a smart water grid and uses smart domestic metering, pressure management, network monitoring and remote leakage detection. Water use minimisation is based on the principle of demand management.

1.4.2.2 Sewerage Systems

Managing the sewerage networks and wastewater treatment works so as to minimise their net energy needs, the best application of assets for transporting and treating wastewater and minimising the environmental impact of the wastewater. This includes managing flows of municipal sewage, industrial effluents, and storm (rain) water and relating these flows to the systems' storage and treatment capacity. Applications include flow metering and network condition monitoring,

1.4.2.3 Energy Use and Recovery

Minimising the amount of energy needed across the water cycle through controlling energy use, optimising power consumption, and by using water and wastewater flows to generate electricity along with recovering energy embedded in the wastewater. This also extends to nutrient and water recovery from wastewater. This involves network, water treatment and wastewater treatment monitoring to minimise the amount of pumping needed, along with treatment chemicals required and optimising treatment processes for water, nutrient and energy recovery.

1.4.2.4 Smart Environment

The use of real-time monitoring allied with predictive systems to minimise the response time to any perturbations in each catchment area, including linking treatment works to the monitoring data. Demand management for municipal, industrial and irrigation applications is used to minimise the amount of water that needs to be abstracted from each catchment area, along with real-time monitoring of water flows through the catchment to maintain the integrity of the water cycle.

1.4.2.5 Flood Management and Mitigation

Real-time monitoring of rainfall, water flows, soil moisture, and groundwater levels are allied to comprehensive and fully updated data on the flood characteristics of each catchment area to respond to changing water levels and to maximise the time available to respond to potential flooding incidents.

1.4.2.6 Resource Management

Monitoring of surface water, reservoir and ground water levels and quality and to ally this data with current and anticipated water demand from various user types to ensure adequate water availability and to balance demand with the various resources available.

1.4.2.7 Integrated Water Management

Smart water systems offer the potential to deliver closed-loop systems for municipal and industrial customers through linking up the treatment, distribution and resource recovery processes outlined in the first six examples. The municipal applications would be based on localised systems, serving a smaller town or a sub-district within a larger utility. This involves distributed rather than centralised treatment facilities with an emphasis on minimising the energy intensity of the water and wastewater networks involved.

1.4.3 Smart Water and the 'Food, Water, Energy, and the Environment Nexus'

Smart water has a central role to play in the so-called food, water, energy and the environment nexus ('the nexus'), especially through water demand management and effluent resource recovery through the nexus. While it is an arresting expression, 'the nexus' may well be replaced by a more compelling expression in time.

Nexus-related themes include irrigation water for agriculture and other applications, nutrient recovery for fertilisers and energy recovery for treatment processes and export. Indirect impacts include lowering water abstraction and utility footprints. There are also direct and indirect interrelationships between resource recovery and the costs associated with maintaining and extending municipal water and sewage services. Water recovery also has an impact on demand and resource management through the impact water reuse on overall water abstraction.

1.5 Water, Smart Water and Cleantech

Water has sometimes had a somewhat uneasy relationship with the rest of the Cleantech sector. This stems from an assumption that pipes, sewers and treatment works do not naturally belong in a sector that is associated with photovoltaics, hydrogen cells and data systems. Such a view does not reflect the fact that water services, gas, telecoms, and electricity provision are utility activities. There are significant cross-linkages between utilities both in terms of services developed for one utility being adapted for another and where combined services can be offered.

As will be seen, water occupies a small section of the Cleantech sector in terms of funding flows and to a lesser extent in capital and operating expenditure, and the same applies with smart water and smart Cleantech. Compared with many sectors where Cleantech is being developed, water and wastewater are seen as slow moving and risk adverse, with some reluctance by municipal and domestic customers to increase up-front spending on their utilities, especially on innovative approaches. The challenge in funding associated with water utilities and services is in fact becoming a driver for water Cleantech in general and smart approaches in particular.

With their relatively small market size, water Cleantech and smart water have tended to be seen as an adjunct for other sectors, where extent technologies can be adapted to extend their market reach rather than looking for approaches that are specifically designed to serve the water sector. Likewise, other utilities and service providers are looking for opportunities in water and wastewater for technologies and techniques that were developed for applications in other sectors.

Other links with the rest of Cleantech are emerging through work on de-carbonising the water and wastewater sector, or making traditionally intensive actions such as water and wastewater pumping and wastewater treatment and recovery energy (and therefore carbon) neutral. Meanwhile, forms of automation are an example of utility services being adopted by other, while smart meters are being developed that combine reading and billing for water and electricity provision.

1.6 Disruption and a Conservative Sector

1.6.1 Why Water Utilities are Risk-Averse

Risk aversity and an institutionally conservative approach are characteristics of the water sector. Unlike for example electricity, water provision is directly affected by public health and environmental concerns. Water is usually expected to meet applicable levels of purity as well as service delivery expectations both in the reliability of water provision and in aesthetics such as taste and colour. Wastewater treatment and disposal is likewise affected by legislation affecting the way it is handled and discharged into the ambient environment.

In developed economies, any deviation from perfect water and wastewater delivery is considered as unacceptable. Shannon and Weaver (1949) pointed out that when information comes across steadily, it is not noticed (background music for example) until it stops. It is the deviation from a steady state that a consumer does not expect. In the case of water and sewerage, any deviation from a perfect service will be immediately apparent and therefore completely unacceptable.

While access to reliable telecommunications services are desirable and humanity can exist without electricity (albeit, with an even greater loss of utility) access to potable water is essential to life, while the economic and public health costs of poor access to water and inadequate sanitation are considerable.

Another factor is the asset intensity of water services – and even more so for sewerage – in relation to the revenues their activities generate. This leads to concerns about stranded assets, whereby innovation obliges a utility to acquire new systems even though it already has perfectly functional assets. For example, if manual read water meters were recently purchased, this may delay the adoption of smart meters because of the concerns about purchasing these assets twice over.

1.6.2 A Question of Standards

Like other utility services, water and wastewater are typically governed by national standards. For water quality, these are led by the World Health Organization's guidelines (WHO, 2011) which are then adopted at the national level, for example the Water Supply (Water Quality) Regulations, 2000 in England and Wales. In Europe, a series of directives also cover water and wastewater standards, including; Drinking Water (1998/83/EC), Bathing Water (1976/160/EEC and revised as 2007/7/EC), Urban Wastewater Treatment (1991/271/EEC) and the Water Framework (2000/60/EC).

Regardless of the power source, electricity will be delivered to a common standard across a utility and indeed a country. Likewise, telecommunications services depend on commonly agreed transmission protocols for both fixed-wire and mobile services. Both services are cheap to transmit over substantial distances between population centres in relation to the revenues these services generate.

In contrast, every water catchment area has its own characteristics. These include the amount of rainfall, its patterns and seasonality, the underlying rocks, geomorphology (the interaction between the landscape and its underlying strata), the presence of aquifers, land use and run-off, population density and distribution, the relation between renewable water resources and demand, and how water and wastewater is managed within the area.

Water is comparatively expensive to transport across catchment areas in relation to the value of the service. Where a utility uses water from a variety of sources, each source may need a specific treatment regimen before it can be released into the distribution network. Indeed, water from different sources will react differently when passed through the mains network (more acidic water will reach with iron pipes, causing corrosion and discolouration) and domestic networks (more acidic, or plumbosolvent water will dissolve lead pipes and solder, which may raise lead concentrations above the applicable standards).

1.6.3 Disruption in a Conservative Sector

A disruptive technology is one which changes the nature of its intended market. For example, railways, internal combustion engines and commercial flight have had a disruptive influence in the business of transportation, as have the telegraph and fixed-wire and mobile telephony in communications.

Despite its conservative nature, significant disruptive events have taken place in the water and sewage sectors. Examples of genuinely disruptive developments in water and wastewater services include the first slow sand filtration system for water for large scale water treatment which opened in Paisley, Scotland in 1804 (Huismann and Wood, 1974) and the development of the activated sludge sewage treatment by Edward Arden and William Locket in 1913–14 (Alleman, 2005).

More recently, reverse osmosis for desalination was developed by Sidney Loeb and Srivasa Sourirajan from the late 1950s and the first commercial reverse osmosis desalination at Coalinga, California entered service in 1965 (Loeb, 2006) and membrane technologies for wastewater treatment and water recovery were transformed by the development of the submerged membrane bioreactor in 1989 (Yamamoto et al., 1989).

Most current and anticipated smart water developments are set to offer incremental rather than disruptive improvements in efficiency and cost-effectiveness. It is the potential ability to integrate and to redouble these incremental benefits into a smart water system that is disruptive.

1.7 The Size of this Market; Estimates and Forecasts

How big is the market for smart water systems and products and how big might it become? A wide number of companies carry out research on the current and forecast size of various technology markets. Data from surveys that are in the public domain (available through press releases, conference presentations or in openly available surveys) is presented in six tables below and placed in its context.

What counts as 'smart' varies from survey to survey as well as the actual amount of hardware involved. Because of the broad nature of definitions used for smart water between the companies surveying the sector there is an equally broad range of market estimates as well as forecasts. No survey is likely to be definitive and no one survey may be more accurate than another, but by comparing them, an overall impression can be made. Their value lies in showing how analysts following the sector perceive its current status and its potential growth and how this perception is changing over time.

The differences between market estimates over time also highlight the relatively early stage of this market's evolution and that it is a sector that is in rapid phase of development. Surveys will vary from year to year due to currency fluctuations against the US dollar as well as changing assumptions about future economic growth.

1.7.1   A Survey of Surveys

A total of 22 surveys and forecasts have been noted, eight covering the overall market (Table 1.1) and 14 looking at specific sub-sectors (Tables 1.2 to 1.6). A CAGR (compound annual growth rate) has been calculated where it was not initially available, to allow the comparison of growth projections. The Lux (2010) survey was one of the earlier ones notes and forecast growth rates are particularly high because of the small market base at the time of the survey.

It should be noted that the market estimates and forecasts provided by Marketsandmarkets are higher in their 2015 and 2016 surveys than in the 2013 survey. A lower CAGR forecast for 2016–21 than for 2015–20 reflects a higher initial market size.

The more recent surveys start from an appreciably higher market estimate base and generally point to a market that will be more substantial than previously anticipated.

Looking at their 2013 market estimates for sub-sectors, GWI (2014) splits the market into four main areas: network optimisation ($726 million), leakage management ($1,494 million), metering and customer services ($1,322 million) and water quality monitoring ($77 million).

Table 1.1 Smart water – overall surveys.

$ billion	Start year	End year	Start	End	CAGR
Lux (2010)	2009	2020	0.50	16.30	37.3%
IDC Energy Insights (2012)	2011	2016	1.40	3.30	18.7%
GWI (2014)	2013	2018	3.62	6.90	13.8%
Marketsandmarkets (2013)	2013	2018	5.43	12.03	17.2%
Transparency (2014)	2012	2019	4.81	15.23	17.9%
Marketsandmarkets (2015)	2015	2020	7.34	18.31	20.1%
Marketsandmarkets (2016a)	2016	2021	8.46	20.10	18.9%
Technavio (2016)	2015	2020	7.00	16.73	19.0%

Adapted from Transparency Market Research (2014); Minnihan (2010); Marketsandmarkets (2013, 2015, and 2016a); IDC (2012) and Global Water Intelligence (2014); Technavio (2016).

Table 1.2 Smart meters.

$ billion	Start year	End year	Start	End	CAGR
TechNavio (2008)	2008	2012	0.24	0.51	20.1%
Lux (2010)	2009	2020	0.21	6.30	36.0%
Pike Research (2011)	2010	2016	0.41	0.86	13.1%
IMS Research (2011)	2010	2016	0.55	0.95	9.5%
Frost+Sullivan (2014)	2013	2017	3.48	5.18	10.5%
IHS Tech (2014)	2013	2020	0.58	1.23	11.5%
Marketsandmarkets (2016b)	2015	2021	3.73	5.67	7.2%
Technavio (2017)	2016	2021	4.83	12.18	20.3%
Research and Markets (2017)	2015	2025	3.75	8.80	8.8%

Adapted from TechNavio (2008 and 2017); Minnihan (2010); Pike Research (2011); IMS Research (2011); Frost and Sullivan (2014); IHS Tech (2014); Marketsandmarkets (2016b) and Technavio (2017).

Table 1.3 Smart water networks.

$ billion	Start Year	End Year	Start	End	CAGR
Lux (2010)	2009	2020	0.16	3.30	31.5%
Navigant Research (2013)	2013	2022	1.12	3.30	12.8%
Frost+Sullivan (2012)	2010	2020	0.35	6.44	33.8%
Navigant Research (2016)	2016	2025	2.50	7.20	11.2%

Adapted from Minnihan (2010), Frost and Sullivan (2012) and Navigant Research (2013 and 2016).

Table 1.4 Leakage management.

$ billion	Start year	End year	Start	End	CAGR
GWI (2014)	2013	2018	1.49	2.80	13.3%

Adapted from Global Water Intelligence (2014).

Table 1.5 Water mapping.

$ billion	Start year	End year	Start	End	CAGR
Lux (2010)	2009	2020	0.02	3.20	56.6%

Adapted from Minnihan (2010).

While the Navigant 2016 CAGR is lower than the 2013 forecast, the anticipated market size is appreciably greater and indeed the 2016 market estimate is almost the size of the previous survey's 2020 forecast.

Table 1.6 Water quality monitoring.

$ billion	Start year	End year	Start	End	CAGR
Lux (2010)	2009	2020	0.11	1.10	23.4%
GWI (2013)	2013	2018	0.08	0.14	13.4%

Adapted from Minnihan (2010) and Global Water Intelligence (2014).

The water testing and analysis market remains dominated by traditional approaches. Marketsandmarkets (2015b) forecasts the overall market, including laboratory systems will be worth $3.5 billion in 2019, growing at 5.2% pa between 2014 and 2019.

According to Aquaspy (Aquaspy, 2013), $210 million was spent on smart irrigation in 2012; $100 million on water irrigation control systems, $30 million on monitoring, $10 million on 'fertigation' (combined fertilisation and drip irrigation systems) and $70 million on greenhouse control systems. Marketsandmarkets (2015c) estimates the overall soil moisture sensing market was worth $98 million in 2015. The smart irrigation market is analysed in greater detail in Chapter 7.

Another way of considering smart water in its broader context is to look at overall spending on smart systems and the hardware that is directly related to it such as metering and monitoring hardware. This was examined by GWI (2016) as 'digital water' with a market with $20 billion in 2014 and projected to grow to $30 billion pa by 2020. The market sizes for treatment and distribution and collection are seen as broadly equal in size.

The main areas of difference between 'digital' and 'smart' relate to those elements of testing, metering and sensing, which while part of smart networks, are not smart appliances in themselves. Parts of automation and control will also include non-smart elements. What these numbers highlight is the non-smart aspects of water hardware that enable smart systems to operate.

To put the smart water figures into their broader context, Marketsandmarkets (2015) estimates that the smart cities market (covering all urban services) was worth $411 billion in 2014 and will grow to $1,135 billion by 2019. GWI (GWM 2015, 2014) estimates that capital spending in 2013 on water infrastructure was $102 billion, rising to $131 billion by 2018 and $110 billion rising to $142 billion for wastewater over the same period.

1.8 Venture Capital Funding Flows

The information in this section is restricted to that which has been provided at conferences, in press releases and articles that are in the public domain. The awkward relationship between water and the rest of the Cleantech sector is highlighted in Venture Capital spending. For example, Boogar Lists is a USA based database of over 2,000 venture capital and Private Equity firms (Boogar Lists, 2014). It lists 89 Cleantech venture capital funds globally, but these do not include Apsara Capital (London) or XPV Capital (Toronto). These are the only dedicated water Cleantech venture capital funds known to be in operation during this period.

Table 1.7 outlines overall venture capital investment in the water sector between 2006 and 2013.

Table 1.7 Water Cleantech venture capital funding, 2006–13.

Water Cleantech	2006–07	2008–09	2010–11	2012–13	2014–15
Number of VC deals	56	159	204	257	139
VC funding ($ million)	293	915	936	1,023	587
Funding per deal ($ million)	5.2	5.8	4.6	4.0	4.2

Adapted from The Cleantech Group (2011) and i3 (2014, 2015 and 2017).

While there has been a decrease in the average deal size since 2009, the overall level of investment has been maintained. However, in 2016, there were 42 water VC investments, generating $173 million in funding. This is the lowest annual figure since 2006 with an average investment size of $4.1 million (i3, 2017).

Placing these figures, Table 1.8 outlines overall Cleantech venture capital investment and the relative size of water investment.

This indicates that the water sector has attracted a consistently small proportion of Cleantech venture capital funding to date.

1.8.1 Smart Water Cleantech Funding

According to the CleanTech Group (Neichin, 2011), in the five years from 2006 to 2010 'smart technology' accounted for 11% of total CleanTech venture capital funding ($3,910 million out of $35,210 million Cleantech funding overall) with 2% of this going on Smart Water ($80 million) compared with $1,600 million on Smart Grid, $686 million on Smart Buildings, $864 million on Smart Transportation and $680 million on Smart Industrial. This does not include indirect investment in smart water through smart grid and smart industrial companies.

1.8.2 Funding Smart Water Companies

With a significant number of smaller, privately owned companies driving much of the sector's development, venture capital is a particularly important element of the business of smart water.

Table 1.8 Cleantech and Water Cleantech venture capital funding, 2006–15.

Overall for Cleantech	2006–07	2008–09	2010–11	2012–13	2014–15
Number of VC deals	1,146	1,574	2,180	2,484	1,965
VC funding ($ billion)	11.4	16.9	20.4	15.9	18.8
Funding per deal ($ million)	10.0	10.7	9.4	6.4	9.6
Water as % of Cleantech funding	2.5%	5.4%	4.6%	6.4%	3.1%
Funding per deal ($ million)	5.2	5.8	4.6	4.0	4.2
Water deal size as % of Cleantech	52%	54%	49%	62%	44%

Adapted from The Cleantech Group (2011); The Cleantech Group (2013); The Cleantech Group (2014); Javier (2011); Haji (2012); Neichin (2011) and Cleantech Group (2016a).

Table 1.9 Evolution of smart water VC funding, 1998–2009.

Funding round	1998–01	2002–05	2006–09
A – Seed/early stage	73%	63%	37%
B – Development	27%	27%	31%
C – Commercialisation	0%	10%	17%
D – Expansion	0%	0%	15%

Adapted from Minnihan (2010).

1.8.3 The Evolution of Venture Capital Funding

Venture capital funding is based on a series of fund raisings, in effect, taking a company from its foundation based upon a potential product or service through to its eventually being listed or acquired. Table 1.9 summarises how funding of smart water companies has evolved over a 12-year period.

In a separate survey, in 2009, 70% of deals were for early stage investments (A and B in Table 1.9), while in 2010-13, the proportion varied between 46% and 55% (i3, 2014a). In 2016 (i3, 2017), 24% of water companies at the VC stage were pre-revenue (seed / early stage), 24% development (revenues of less than $500,000), 29% at commercialisation (revenues of $0.5-5.0 million) and 23% at the expansion stage (revenues above $5.0 million).

While there is an evident shift from early to later stage investment during this period, this change is not as straightforward as it may appear. Since the economic and financial sector turndown in 2008 venture capitalists have concentrated on development and expansion capital rather than seed and early stage capital, because while it offers the prospect of lower returns, it is also a lower risk investment.

Venture capital investors who are concerned about the conservative nature of the water sector do not appreciate the nature of the sector; 'in all industries…there is no such thing as a non-conservative customer' (David Henderson, comment at the World Water Tech Investment Summit, London, 29th February 2012).

Henderson also observed that there are enough companies looking for funding but the degree of entrepreneurial talent (the ability to commercialise and sell their offerings) in these companies is notably weak compared with their engineering capabilities. This means that more mentoring and support is needed than is usually the case. This is a concern as there are in the region of five venture capital funds worldwide which have four or more investments in water companies, as far as water Cleantech is concerned, the venture capital sector as a whole lacks the management support, market understanding and thematic commitment that is needed by these companies.

1.9 Two Perspectives on Venture Capital and New Technologies

As far as investors are concerned, water Cleantech investments have been characterised by a notably long period between the initial investment and the point when these

Table 1.10 Water companies in the Global Cleantech 100.

Number of entries in top 100	2010	2011	2012	2013	2014	2015	2017
All water companies	10	11	10	10	13	11	9
Smart water companies	3	2	2	1	3	2	2

Adapted from Cleantech Group, 2011, 2012, 2013, 2014, 2015, 2016b and 2017.

investments can be realised. It may be that this not so much the case when it comes to smart water Cleantech, but investor caution remains.

1.9.1 The Global Cleantech 100 – Cleantech Companies to Watch

Since 2010, The Cleantech Group has invited a panel of judges to select the top 100 emerging Cleantech companies (Table 1.10). It provides an annual snapshot into how these judges perceive innovation in the Cleantech sector to be developing. It also provides a snapshot as to their perception of water and smart water companies within this top 100.

Companies listed are involved in for smart irrigation (AquaSpy), smart metering (Fathom, three appearances), pressure management (i2O with three appearances), leakage management (Takadu, with six appearances) and water quality monitoring (Universtar).

This suggests that emerging water Cleantech companies are consistently more popular as prospects than as investment targets and that appears to be especially the case when considering smart water Cleantech companies.

1.9.2 The Gartner Hype Cycle – Investor and Customer Expectations and Realities

Gartner is a Stamford, CT based information technology consulting firm founded in 1979.

Through its experience in following for example the development of mobile communications in the 1980s and the Internet in the 1990s, Gartner has examined new technologies and their application are perceived by investors and customers. To this end, they developed the Hype Cycle, a tool for looking at the commercial evolution of an emerging technology. The Gartner Hype Cycle (Gartner, 2015a) consists of five stages. The interpretations in Table 1.11 are the author's.

Table 1.11 The Gartner Hype Cycle and smart water (1 = low to 5 = high).

	Trigger	Peak	Trough	Slope	Plateau
Risk	5	4	3	2	1
Opportunity	5	3	4	3	3
Funding need	1	2	4	5	3

Author's analysis, using the Gartner (2015a) framework.

1) Technology trigger (on the rise) – This where a prototype (potential new product) is announced, which has potential applications (proof of concept) but have not been tested in real-life conditions. The work is funded by personal investment or company funding, grants, angel investors, family office investment, and sometimes by early stage venture capital (first round of VC fund raising).

2) Peak of inflated expectations (at the peak) – pre-commercial trials of the prototype start, some of which may work, while others fail. This is accompanied by considerable media interest, looking at the possibilities for the technology. Initial contact with potential clients is made and external investment shifts towards family office funding early stage venture capital (first and second VC fund raising rounds).

3) Trough of disillusionment (into the trough) – Delays in product development and problems replicating lab bench performance in real-life conditions mean that the failure rate reaches its peak. Many projects are abandoned and those which continue need continued support from the company and further funding to develop applications that their potential clients need through pilot tests (first to third stage VC fund raising rounds).

4) Slope of enlightenment (climbing the slope) – Commercially viable applications are demonstrated through pilot tests and early adopter customers. Further product development widens the product's applicability and performance. Some products are acquired by incumbent companies, with funding coming from late stage venture capital (second to fourth VC funding rounds) and expansion capital.

5) Plateau of productivity (entering the plateau) – The product is increasingly seen as commercially and technically viable and is integrated into extant systems and applications. The emphasis shifts towards commercialisation, marketing and developing new versions and applications. The product is mainly funded by late stage venture capital (second to fourth VC funding rounds) and expansion capital with a trade sale to an incumbent player or a Stock Exchange Listing under consideration.

A tool such as the Hype Cycle allows companies and investors to consider how technologies are involved from the perspective of broad market acceptance, rather than that of enthusiasts such as early adopter utilities or early stage investors.

During 2015, Gartner published a series of updates (Gartner, 2015b, 2015c, 2015d, 2015e and 2015f) looking at the development of various technologies. In 2015, water management was seen as just entering the trough of disillusionment (stage 3), reaching the plateau in 2–5 years, having been noted at the peak of inflated expectations in 2012 (2–5 years to the plateau).

Other relevant areas seen at the peak (stage 2) in 2015 include; energy water nexus, smart city framework, sustainable performance management, the Internet of Things, meter data analytics and big data in energy and utilities. In contrast, geospatial imagery for utilities and geospatial platforms are at the technology trigger stage (stage 1). Other themes seen as going into the trough include advanced metering infrastructure, asset performance management and asset investment planning tools while climbing the slope to the plateau (stages 4 and 5) includes meter data management and environmental monitoring and control.

It is evident that with the exception of meter data management and environmental monitoring and control, most smart water applications are perceived as emerging technologies.

1.10 Sales of Smart Systems

The Gartner Hype Cycle (1.9.2) is a useful reminder that smart water is, and will be for some years, an emerging technology. Even so, significant commercial sales are being seen in a number of areas, as will be discussed later in this book. Smart water metering is the most obvious case, with utility wide rollouts seen for example in Malta, England (Southern Water), and the USA (Global Water).

The market estimates in section 1.7 indicate sales of smart water products in the region of $3.62–7.34 billion in 2013–15. The principal areas of activity, along with water meters, are smart networks, leakage management and smart irrigation. It is a market which is being primarily driven by early adopters, utilities, industrial water users and growers who stand to benefit the most from approaches that have yet to be fully commercially adopted.

1.11 Smart Water for Consumers

In terms of public perception, this is by some way the largest sector to date. Even so, the development of customer interfaces is at an early stage in terms of being easily used and manipulated and the main challenge there is to influence and inform customer behaviour over a significant period of time.

At present, the emphasis is on modifying consumer behaviour through helping them to minimise their water use. The next step is going to be in enabling them to notice internal leaks and in the timing of their water use to ease peak demand. Due to the amount of energy consumed when heating water, consumers can also be alerted to potential energy savings.

1.12 Smart Water for Utilities and Industrial Customers

A broad suite of applications are being used or are under development for municipal and industrial customers. Smart systems are either installed when replacing or upgrading extant systems (metering and monitoring for example) or as entirely new offerings. For example, smart meters are being installed by a number of utilities in the UK where customers previously did not have water meters. Smart approaches are being used to minimise the costs of providing services through optimising the efficiency of extant assets and by preventing the development of surplus assets through the more efficient use of extant assets and suitable demand management measures. A secondary attraction for many of these customers lies in the scope for integrating a number of incremental improvements. This means that interoperability between systems and common standards are a particular concern.

Utilities currently using smart water approaches include Hera-Modena (Modena in Italy, remote reading systems for real-time smart water meters and other urban utility services), Haghion (Jerusalem in Israel, real-time network monitoring for early leakage detection), Wessex Water (UK, trials for smart sewer metering), Aguas de Cascais (Portugal, integrated non-revenue water reduction), South East Water (UK, mains water pressure monitoring and management), Northumbrian Water (UK, integrated data access and control at the regional level), and Vitens (Netherlands, remote and real-time water quality monitoring).

1.13 Irrigation and Surface Water Monitoring

Software systems for the remote monitoring of soil moisture enable water and nutrients to be introduced in the most effective manner, improving yield as well as lowering water consumption. Smart irrigation systems can also optimise the effective delivery of water and nutrients. These also apply for urban landscape irrigation software to minimise park, playing field and garden watering, in turn lowering water use, nutrient run-off and erosion. These will be discussed in more detail in Chapter 7.

Smart water is also being used in reservoir and dam management, and monitoring of surface water quality and inland water flows and modelling and monitoring flood vulnerability. Increasing rainfall variability as a result of climate change is a particular driver behind the need for more rapid monitoring of surface water quality and flow and the ability to predict flood and drought events as far ahead as possible.

1.14 Water and the 'Internet of Things'

The Internet of Things refers to the interconnection of monitoring systems into what is seen as an all-encompassing whole and therefore it is a somewhat poetic expression for imagining a world which is limitlessly and universally interconnected via the Internet. It comes in part from the idea of 'big data' that the analytical power of information increases as more data from more sources is integrated. The 'IoT' expression was coined by Kevin Ashton in 1999 (Ashton, 2009). It functions via device to device communications where each device can be identified within a network so as to provide a coherent and comprehensive coverage of the data sought. A particular area of interest is in integrated monitoring and management of domestic appliances.

For the water sector, the Internet of Things is an extension of the smart water concept to include other applicable and related services, thereby linking water with other utilities and water consuming devices. The Internet of Things has the potential to alter popular expectations about utility services, water consumers, and how these affect other areas. For example, bathing water quality data in real-time becomes a consumer tool while the role of irrigation agriculture can be transformed when it enables optimal water and nutrient inputs to be deployed.

1.15 Some Initial Caveats

Chapter 9 will discuss in some detail some of the broader challenges facing smart water. The two points considered here deserve an immediate mention because they will recur through the rest of this book.

1.15.1 A Caveat about a Swiftly Evolving Future

This survey is focussed on technologies and techniques which are either in development or have only recently been deployed. It is therefore biased towards possibilities rather than potential pitfalls. We only have forecasts for the scale and speed of the deployment of these new approaches.

Customer behaviour is often overlooked by utilities. Likewise, the deployment of a large array of low cost, Internet-linked domestic devices as part of the Internet of Things carries the risk that their system security and integrity will be lower than for more expensive appliances and applications. They may become a 'back door' for accessing customer information or interfering with operations.

Smart water systems are constantly emerging and evolving. For example, during the UK floods in December 2015 to January 2016, drones were used to gather data about river flows and flood plain inundation at pressure points down river systems, covering 60 km of river bank in a day (Kinver, 2016). This allowed for the first time an insight into how flood plains were in fact holding water during the floods.

1.15.2 A Caveat on Data and the Silo Mentality

Unless concentrating on a deliberately small set of criteria, it is essential to ensure that all possible data sources can be accessed and usefully integrated. Gillian Tett's book 'The Silo Effect' (Tett, 2015) contains a telling anecdote about how open data allowed New York City to deal with yellow (catering) fat in its sewers. New York's government has been traditionally run on compartmentalised lines and communications between its various activities has been an impediment in addressing chronic problems affecting the city such as fire risk and fraud. In 2011, a team was recruited with a remit to use all of the city's data sources to address such problems. To identify where the fat was being generated and discharged, they looked at sewer blockage incidents and related them to tax returns, business licences and kitchen fires. From this, they identified the most effective data came from listing kitchens which had not applied for the appropriate waste disposal licence. Instead of threatening these businesses with prosecutions, the team presented the database to the health and safety and fire inspection departments and another department that was seeking to promote biofuel recycling. When the caterers were advised that they were throwing a valuable resource away, compliance became commonplace.

There is a danger of extrapolating too much from limited information when considering an emergent application such as smart water. Chapter 9 will consider at some length all the caveats and concerns that have been identified to date.

Conclusions

Smart water is an emerging subset in Cleantech and smart Cleantech in particular. In contrast with the rest of the Cleantech sector, it involves serving a notably conservative and risk-adverse business and one which faces continuing funding shortages.

The challenge for smart approaches is to assist clients in delivering more and better services at a lower cost than before by enabling them to adopt more radical approaches than they would hitherto have countenanced. For the water sector, this also means improved higher degrees of public health and environmental compliance at a swifter pace than traditionally seen.

As industrial clients do not face the same set of constraints as utilities, but share the same needs to improve the water-related efficiency of their processes, they can serve as a platform for commercialising innovative approaches.

The conservative nature of the water sector means that the technology which will be widely adopted in 2035 is already here (Sedlack, 2016) as water technology typically takes 20 years to move from conceptual development to broad adoption (Parker, 2011). The Gartner Hype Cycles suggest shorter gestation periods, based on experiences in less conservative sectors. It may be that one of the distinguishing features of smart water will be a more rapid adoption process than the sector normally experiences. Nevertheless, the basic elements of a broad variety of smart water approaches are already being developed and applied at the commercial level. While the essentials may remain relatively unchanged, quite a lot may happen in the way that smart water approaches are developed, adapted and adopted.

In commercial terms, smart water appears to be a small market. That would be misleading as much of smart water's potential lies in being a tool for improving the effective deployment of other assets and avoiding the need to invest as much in new assets as originally planned.

To some extent, smart water and water and the Internet of Things are more of a philosophy than a single concept (Reynolds, L. personal communication, August 2016). Both have been enabled by the emergence of technological innovation and user-centred applications and their application across the business of water services. It will become evident that smart domestic applications (especially those using the Internet of Things) are quite separate from smart approaches designed to optimise water and wastewater network efficiency and that smart irrigation is quite distinct from both of these. What unites them is their potential to use water and wastewater more efficiently while lowering the cost of their delivery and treatment and enabling operators and consumers to be fully informed about the impact of their water usage and to appreciate how their behaviour can be modified towards more sustainable consumption patterns.

These smart water approaches have three chief elements in common; the potential to use water and wastewater more efficiently, lowering the cost of their delivery and treatment and informing operators and consumers. Informed operators and consumers appreciate the cost and impact of their water usage and consider how their behaviour can be modified towards more sustainable consumption patterns.

References

Alleman J E (2005) The Genesis and Evolution of Activate Sludge Technology, http://www.elmhurst.org/DocumentView.aspx?DID=301 accessed 20th May 2011.

Anderson R and Fuloria S (2010) On the security economics of electricity metering. Paper presented to the 9th Workshop on the Economics of Information Security (WEIS 2010), Harvard University, USA, June 8th 2010.

Aquaspy (2013) Presentation to the World Water Tech Investment Summit, London 6–7 March 2013.

Ashton (2009) That 'Internet of Things' thing. RFiD Journal, 22nd June 2009. http://www.rfidjournal.com/articles/view?4986.

Boogar Lists (2014) http://www.boogar.com/resources/venturecapital/clean_tech2.htm.

Cleantech Group (2011) Global Cleantech 100 2010. Cleantech Group LLC, San Jose, USA.

Cleantech Group (2012) Global Cleantech 100 2011. Cleantech Group LLC, San Jose, USA.

Cleantech Group (2013) Global Cleantech 100 2012. Cleantech Group LLC, San Jose, USA.

Cleantech Group (2014) Global Cleantech 100 2013. Cleantech Group LLC, San Jose, USA.

Cleantech Group (2015) Global Cleantech 100 2014. Cleantech Group LLC, San Jose, USA.

Cleantech Group (2016a) Global Cleantech 100 2015. Cleantech Group LLC, San Jose, USA.

Cleantech Group (2016b) Q4 FY/2015 Innovation Monitor. Cleantech Group LLC, San Jose, USA.

Cleantech Group (2017) Global Cleantech 100 2016. Cleantech Group LLC, San Jose, USA.

Drago C M (2009) The Smart Grids in Italy - an example of successful implementation. Presentation by IBM to the Polish Parliament, October 27th 2009. http://www.piio.pl/dok/20091027_The_Smart_Grids_in_Italy_-_an_example_of_successful_implementation.pdf.

European Union (2009) Directive 2009/72/EC concerning common rules for the internal market in electricity and repealing Directive 2003/54/EC. Official Journal of the European Union, August 14th 2009, L 211/55–93.

Forer G and Staub C (2013) The US water sector on the verge of transformation. Global Cleantech Centre white paper. EY, New York, USA.

Frost and Sullivan (2012) Global Smart water grids to see increasing investment over the next decade. Presentation to the World Water Tech Investment Summit, London, 29th February 2012.

Gartner (2015a) Hype Cycle Methodology. http://www.gartner.com/technology/research/methodologies/hype-cycle.jsp.

Gartner (2015b) Gartner Hype Cycle for Utility Industry IT, 2015. https://www.gartner.com/doc/3096022?ref=ddisp.

Gartner (2015c) Hype Cycle for Smart Grid Technologies, 2015. https://www.gartner.com/doc/3092617?ref=ddisp.

Gartner (2015d) Hype Cycle for Green IT, 2015. https://www.gartner.com/doc/3101522?ref=ddisp.

Gartner (2015e) Hype Cycle for Sustainability, 2015. https://www.gartner.com/doc/3101519?ref=ddisp.

Gartner (2015f) Hype Cycle for the Internet of Things, 2015. https://www.gartner.com/doc/3098434?ref=ddisp.

Global Water Intelligence (2014) Market Profile: Smart water networks. Global Water Intelligence, January 2014, pages 39–41.

Global Water Intelligence (2016) Digital transformations set to boost buoyant smart water market. Global Water Intelligence, October 2014, pages 35–37.

Gohring N (2008) Microsoft to discontinue MSN Direct. PC World, October 28, 2008. http://www.pcworld.com/article/174581/article.html.

Gómez-Expósito (2012) Smart Grids: so old, so new. Mondragón, December 19, 2012.

GWM 2015 (2014) Global Water Markets 2015, Media Analytics Limited, Oxford, UK.

GWI (2016) Water's Digital Future: The outlook for monitoring, control and data management systems. Global Water Intelligence, Oxford, UK.

Haji S (2012) Quoted in Water Acquisitions Rise: Will Venture Capital Follow? Forbes 28th February 2012.

Heath J (2015) Smart Networks. Presentation to Smart Water: The time of now! SWAN Forum Conference, London April 29–30th 2015.

Henderson D (2012) Comment at the World Water Tech Investment Summit, London, 29th February 2012.

Huismann L and Wood E (1974) Slow sand filtration. World Health Organization, Geneva, Switzerland.

i3 (2014a) i3 Quarterly Innovation Monitor: Water and Wastewater 4Q 2013. Cleantech Group LLP. Page 2.

i3 (2014b) i3 Quarterly Innovation Monitor: Water and Wastewater 3Q 2014. Cleantech Group LLP. Page 1.

i3 (2014c) i3 Quarterly Innovation Monitor: Water and Wastewater 4Q 2013. Cleantech Group LLP. Page 1.

i3 (2014d) i3 Quarterly Innovation Monitor: Water and Wastewater 4Q 2013. Cleantech Group LLP. Page 2.

i3 (2015) i3 Quarterly Innovation Monitor: Water and Wastewater 4Q 2014. Cleantech Group LLP. Page 2.

i3 (2017) https://i3connect.com/tags/water-wastewater/1181/activity.

IDC Energy Insights (2012) smart water spending forecast: 2012. IDC, Milan, Italy.

IHS Technology (2014) Water Meters Report – 2014. Smart Water Meters Intelligence Service. IHS Markit, London, UK.

IMS Research (2011) The World Market for Water Meters – 2011. IMS Research, Wellingborough, UK.

IMS Research (2013) Smart Water Management Market by Solutions (Network Monitoring, Pressure Management, Analytics, Meter Data Management), by Services (Valve and Information Management, Pipeline Assessment), by Smart Meter Types, by Region – Global Forecast to 2020. IMS Research, Wellingborough, UK.

Javier M (2011) Ontario Global water Leadership Takeaways, Cleantech Group 2011.

Kinver M (2016) UK floods: Drone provides researchers with unique data. BBC, 20-1-2016, http://www.bbc.co.uk/news/science-environment-35353869.

Loeb S (2006) Personal notes on the development of the Cellulose Acetate Membrane, prepared by and delivered by Sid Loeb at an honour ceremony of the first Inaugural Sydney Loeb Award by the European Desalination Association.

Marketsandmarkets (2013) Smart Water Management Market – Smart Water Meters, EAM, Smart Water Networks, Analytics, Advanced Pressure Management, MDM, SCADA, Smart Irrigation Management, Services – Worldwide Market Forecasts and Analysis (2013–2018).

Marketsandmarkets (2015a) Smart Cities Market by Smart Home, intelligent Building Automation, Energy Management, Smart Healthcare, Smart Education, Smart Water, Smart Transportation, Smart Security, and by Services – Worldwide Market Forecasts and Analysis (2014–2019).

Marketsandmarkets (2015b) Water Testing and Analysis Market by Product (TOC, PH, DO, Conductivity, Turbidity), Product Type (Portable, Handheld, Benchtop), Application (Laboratory, Industrial, Environmental, Government) and Region – Global Trends and Forecast to 2019.

Marketsandmarkets (2015c) Soil Moisture Sensor Market by Type (Volumetric and Water Potential), Application (Agriculture, Residential, Landscaping, Sports Turf, Weather Forecasting, Forestry, Research Studies and Construction), and Geography – Global Trends and Forecast to 2020

Marketsandmarkets (2016a) Smart Water Management Market by Advanced Water Meters (Meter Type and Meter Read Technology), Solution (Network Monitoring, Advanced Pressure Management, SCADA System, Advanced Analytics, Residential Water Efficiency), Service – Global Forecast to 2021

Marketsandmarkets (2016b) Smart Water Metering Market by Type, by Component, by Application and by Region – Global Trends and Forecast to 2021.

Michael W (2011) Insight of the week: Smart water investment just a drop in the bucket. Cleantech Insights blog, 28th January 2011, Cleantech Group LLP.

Micro Irrigation System Trends and Global Forecasts (2011–2016) Markets and Markets, March 2012, FB 1772.

Microsoft (2002) Microsoft Launches Smart Personal Object Technology Initiative. http://news.microsoft.com/2002/11/17/microsoft-launches-smart-personal-object-technology-initiative/

Minnihan S (2010) Water IT: How the $16 billion market will take shape. Lux Research. Presentation, 21 September 2010.

Navigant Research (2013) Smart Water Networks. Smart Water Meters, Communications Infrastructure, Network Monitoring and Automation Technologies, and Data Management and Analytics: Market Analysis and Forecasts. Navigant Consulting Inc.

Navigant Research (2016) Smart Water Networks. Water Meters, Communications Infrastructure, and Data Analytics for Water Distribution Networks: Global Market Analysis and Forecasts Navigant Consulting Inc.

Neichin G (2011) The State of Smart Water: Context, Capital and Competition. Presentation by the Cleantech Group to the SWAN Conference, London, 17th May 2011.

Northeast Group, (2015) Global Smart Water Infrastructure: Market Forecast (2015–2025) October 2015.

OECD (2012) Policies to support smart water systems. Lessons learnt from countries experience. ENV/EPOC/WPBWE(2012)6. OECD, Paris, France.

Parker D S (2011) Introduction of New Process Technology into the Wastewater Treatment Sector. Water Environment Research 83: 483–97.

Peleg A (2015) Water Networks 2.0: The Power of Big (Wet) Data. Presentation to Smart Water: The time of now! SWAN Forum Conference, London April 29–30th 2015.

Pike Research (2011) Smart Water Meters: Global Outlook for Utility AMI and AMR Deployments: Market Analysis, Case Studies, and Forecasts.

Research and Markets (2017) Global Smart Water Metering Market Analysis and Trends – Industry Forecast to 2025. Research and Markets, Dublin, Ireland.

Royan F (2014) Global Smart Water Metering Market. Presentation to SMi Smart Water Systems Conference, London April 28–29th 2014.

Sedlack D L (2016) The Limits of the Water Technology Revolution. Presentation to the NUS Water Megatrends Workshop, 25th February 2016, NUS, Singapore.

Shannon C E and Weaver W (1949) The Mathematical Theory of Information. University of Illinois Press, Urbana, USA.

Smart Grid Forum (2014) Smart Grid Vision and Routemap. Department of Energy and Climate Change and Ofgem, London, UK.

Technavio (2008) Global Smart Water Meters Market 2008–2012 forecasts the size of the global smart water meters market over the period 2008–2012. Technavio, London, UK.

Technavio (2017a) 2016–2020 Global Smart Water Management Report. Technavio, London, UK.

Technavio (2017b) Global Smart Water Meter Market, 2017–2021. Technavio, London, UK.

Tett G. (2015) The Silo Effect. Simon and Schuster, London, UK.

The Cleantech Group (2011) Investment Monitor Report.

The Cleantech Group (2013), Press Release, 3rd January 2013.

The Cleantech Group (2014), Press Release, 8th January 2014.

Transparency Market Research (2014) Smart Water Management Market (Component Types – Hardware, Solutions, Services; Meter Read Technology – Fixed network, Cellular network) – Global Industry Analysis, Size, Share, Growth, Trends, and Forecast 2013–2019.

Water Cleantech VC investment in the USA fell from $123 million in 2007 to $67 million in 2009, and recovered to $95-125 million pa in 2010–12.

World Health Organization (2011) Guidelines for Drinking-water Quality, Fourth Edition, WHO, Geneva, Switzerland.

Yamamoto K, Hiasa H, Talat M and Matsuo T (1989) Direct solid liquid separation using hollow fiber membranes in activated sludge aeration tank. Water Science and Technology 21: 43–54.

2

Why do we Need Smart Water?

'When you cannot measure it, when you cannot express it in numbers, your knowledge is of a meagre and unsatisfactory kind.' (Thomson, 1889)

Introduction

If sustainable supplies of water were widely and freely available and utilities provided a universal, safe and affordable water and sewerage service, there would be little need for smart water, other than to further utility efficiency as and when extant assets need to be replaced. Unfortunately, it is evident that this is not the case. Utilities already face considerable difficulties in delivering their services in a satisfactory manner while the challenges of population growth and urbanisation, climate change and dealing with an ageing infrastructure will profoundly exacerbate these challenges.

2.1   The Water Supply Crunch

2.1.1   Water Scarcity and Stress

Water stress is defined as internally renewable water resources of 1,000 to 1,700 m^3 per person per annum and absolute scarcity as being below 1,000 m^3 per annum, with a recently adopted extreme scarcity category at below 500 m^3 per annum (Falkenmark and Lindh, 1976; Falkenmark and Lindh, 1993). According to the United Nations, water stress occurs when more than 10% of renewable freshwater resources are consumed. The European Environment Agency sees water stress starting at 20% of renewable resources being abstracted annually rising to severe stress when abstraction exceeds 40% (EA/ NRW, 2013). Countries with severe water stress often have to rely on non-renewable groundwater supplies or desalination. Increasingly, water reuse is also being adopted.

Smart Water Technologies and Techniques: Data Capture and Analysis for Sustainable Water Management, First Edition. David A. Lloyd Owen.

2.1.2 Renewable Water Resources

Renewable surface water resources are estimated at 42,600 km^3 per annum (range 33,500 km^3 to 47,000 km^3 per annum), along with a renewable groundwater recharge of 2,200 km^3 per annum. These can vary by 15–25% with for example drier years worldwide in 1965-68 and 1977–79 and wetter years in 1949–52 and 1973–75 (Gleick, 1993).

Out of the 42,600 km^3 annual river flows, 20,426 km^3 is lost as surface run-off in floods, 7,774 km^3 flows through remote rivers, chiefly inaccessible parts of the Amazon and Congo, and in northern Europe and America. This leaves a net year-long, usable and accessible water input (basic run-off) of approximately 14,100 km^3 (Postel, Gretchen and Ehrlich, 1996). This compares with an annual abstraction of 3,829 km^3 in 2000 (Molden, 2007). This run-off may neither take place in the right place nor at the right time of year. According to the UN Food and Agriculture Organization (Molden, 2007), 73.4% of water withdrawals in 2000 were from surface waters, against 19.0% from groundwater, 4.8% from drainage water, 2.4% from wastewater reuse and 0.3% via desalination. This means that 2,810 km^3 of surface water was withdrawn or 20% of accessible resources. Surface water flows can also be seasonal; in Asia, 80% of runoff takes place between May and October.

2.1.3 Population Growth and Urbanisation

The world's population is set to rise from an estimated 7.3 billion in 2015 to 9.7 billion by 2050 and 11.2 billion by 2100 (UN DESA, 2015). The population range forecasted for 2100 is between 9.5 and 13.3 billion (UN DESA, 2015). The percentage living in urban areas is set to rise from 53.6% in 2014 to 66.4% by 2050 (UN DESA, 2014). Humanity is encountering the most dramatic process of urbanisation in history. This is taking place worldwide and in all types of towns and cities. It is also happening at its greatest rate and scale where people are the most poorly placed to meet the challenges urbanisation brings.

Population estimates and projections drawn up by the United Nations highlight the scale and trajectory of urbanisation (UN DESA, 2014), notably in Africa and Asia. Over a century, the urban population of Africa is forecast to grow 37-fold, while there will be 3.1 billion more urban dwellers in Asia. Africa's urban population grew by 433 million in 65 years from 1950. It is projected to grow by a further 867 million in the next 35 years. As a result, Asia and Africa's share of the world's urban population is set to increase from 35% to 74% during this period while Europe and North America's falls from 47% to 15%.

Urban living has become the norm (Table 2.1), from 29% in 1950 (53% in developed countries against 18% in developing economies) to 50% by 2010 (75% and 45% respectively) to 69% forecast for 2050 (86% for developed and 66% for developing economies). The nature of urbanisation is also changing. Water and sanitation infrastructure are one of the chief constraints facing the development of mega cities (cities with at least 10 million people). None of the cities forecast to become mega cities in the next few decades have adequate water or sewerage services.

While the mega cities enjoy the most attention, the growth in cities of 1–5 million is equally significant, while growth is slower in secondary cities (Table 2.2).

A move to smaller households also impacts water demand. Research by the Energy Savings Trust (McCombie, 2014, 2015) examined domestic water usage at 86,000 UK

Table 2.1 A century of change – urbanisation, 1950–2050.

Urban population (million)	1950	2015	2030	2050
Africa	39	472	770	1,339
Asia	245	2,113	2,752	3,313
Europe	283	547	567	581
Latin America	69	503	595	674
North America	110	295	340	390
Oceania	8	28	34	42

Adapted from UN DESA (2014).

Table 2.2 Population by city size.

Urban population by size (million people)	1950	2015	2030
10 million or more	23	471	730
5 million to 10 million	32	307	434
1 million to 5 million	128	847	1,128
500,000 to 1 million	65	371	509
300,000 to 500,000	50	262	319
Less than 300,000	447	1,669	1,938

Adapted from UN DESA (2014).

households (Table 2.3). An inverse relationship between water usage and household size was noted. A significant proportion of the difference lies in how domestic appliances are loaded, while smaller households may also have more time for longer showers. This is covered in more detail in Chapter 4.6.2.

Urbanisation increases local water demand, sometimes beyond what can be locally supplied. A limited number of people have been living in water stressed areas for the past 2,000 years (Kummu et al., 2010). Widespread water stress is a more recent phe-

Table 2.3 Water consumption by household size (litres per capita per day).

Household	Per person	Per household
1	154	154
2	143	285
3	140	421
4	134	534
5	128	641
6+	123	813

Adapted from McCombie (2014).

nomenon as outlined in Table 2.1. Between 2010, a majority of people lived in both urban areas and areas of water stress for the first time.

2.1.4 Water Shortage, Scarcity and Stress

There is also a linkage between water stress and development. Where economic development is growing the fastest, water stress is most apparent. Table 2.4 outlines the linkage between the share of global water and economic activity.

Population and economic growth is set to take place in more water scarce areas (Veolia Water, 2011), although the average GDP per capita remains higher in areas with lower levels of water stress (Table 2.5).

While 32% of people living in OECD countries had no or low water stress in 2005 (OECD, 2008) against 37% for the BRIC countries (Brazil, Russia, India and China), 30% of those in the OECD are forecast to experience low or no stress by 2030 against 20% for the BRICs. In contrast, severe stress in the OECD is forecast to rise from 35% to 38%, while rising in the BRICs from 56% to 62% during this period.

The UN Food and Agriculture Organization (FAO) predicts significant renewable water resource shortfalls by 2030 on the basis of 4,200 km^3 of readily accessible annual water flows (Table 2.6). This may be met in part by desalination, water reuse or the abstraction of non-renewable groundwater resources. The latter cannot be seen as a long-term option.

Table 2.4 People living with water shortage, total and percentage of the global population and by m^3 per capita.

Year	Total affected (million people)	Extreme <500	Scarcity 500–1,000	Stress 1,000–1,700	All <1,700
1900	131	0%	2%	7%	9%
1980	1,679	5%	11%	22%	38%
2005	3,247	10%	25%	15%	50%

Adapted from Kummu et al. (2010).

Table 2.5 Water stress and economic development.

Per cent of renewable water withdrawn	% Of global population 2010	2050	% Of global GDP 2010	2050	GDP/population 2010	2050
0–20%	46%	32%	59%	30%	1.28	0.94
20–40%	18%	16%	16%	25%	0.87	1.56
>40%	36%	52%	22%	45%	0.61	0.87

Adapted from Veolia Water (2011).

Table 2.6 Water demand in 2000 and 2030 and the forecast supply shortfall.

Water abstraction (km³ pa)	2000	2030	Shortfall by 2030
Municipal	434	900	429
Industrial	733	1,500	703
Irrigation	2,699	4,500	1,566
Total	3,856	6,900	2,698

Adapted from Molden (2007); FAO (2010).

Making good such a shortfall through desalination and water reuse is in theory possible, but it would carry a significant cost. Assuming all municipal and industrial water could be reused (with the necessary sewerage and sewage treatment infrastructure already be in place, see Table 2.7), this would cost an extra $0.20–0.30/m³, with desalination being needed for municipal water where effluents are inadequately treated. Desalination costs an additional $0.45–0.60/m³ and a further $0.10–0.20/m³ in transport costs for the 40% of humanity living more than 100 km from a coast. If all recoverable water goes to industrial and municipal customers, then agriculture would have to depend on desalination, which does not appear to be a realistic option.

Wastewater can be used for irrigation, but municipal and industrial users would then depend more on desalination. This underlines the resource competition taking place between municipal, agricultural and industrial demand.

2.1.5 Population and Water Stress

Use is driven by demand for municipal water (domestic, municipal and commercial), industrial water (including for energy generation) and irrigation (food and fibre). These are in turn driven by population change and economic development. Water demand for food production is driven by intensive farming, which is usually more water intensive than traditional approaches.

There is a clear link between population growth and water stress, as renewable water resources are finite and can vary over time, while the global population and degree of urbanisation are both increasing. Urbanisation results in higher use of water where people have access to piped water supplies and through economic development gain access to consumer goods ranging from washing machines to baths and even swimming pools.

Table 2.7 Replacing water provision shortfalls via desalination and wastewater recovery in 2030.

Municipal water shortfall (429 km³)	$86–129 bn pa
Industrial water shortfall (703 km³)	$141–211 bn pa
Agricultural water shortfall (1,566 km³)	$767–1,645 bn pa
Total	$894–1,985 bn pa

Author's projections, using UN FAO water shortfall data.

The UN Sustainable Development Goals for 2030 will also increase water consumption as an unintended consequence. The World Health Organization (WHO) defines basic access as less than 20 litres per capita per day from a remote source, intermediate access at 50 litres from a nearby source and optimal access as more than 100 litres from a continual household source (Howard and Bartram, 2003). The 2015 United Nations Sustainable Development Goal 6.1 seeks by 2030 to 'achieve universal and equitable access to safe and affordable drinking water for all' (United Nations, 2015). Safe drinking water means that it meets the World Health Organization's 2011 guidelines for drinking water quality. Increased connections to sewerage networks will have an impact as traditional urban sewerage requires a minimum water input to flush solids through the network to the mains and their final destination.

The scale of the change in water use implied here is shown by the current level of 'safe' drinking water access. Until 2017, the Joint Monitoring Project (JMP) of the UN and WHO used access to 'improved' water and sanitation as their benchmark. From 2017, this has been changed to access to 'safe' water and sanitation (JMP, 2017a). According to the UN/WHO (JMP, 2015) 98% of people in developed countries have household piped water supplies, against 72% in developing countries and 32% in the least developed countries. In 2015, 82% of people have access to 'improved' sanitation and 96% 'improved' drinking water in urban areas and 68% to 'improved' sanitation and 91% 'improved' drinking water globally. An analysis of the 2010 JMP highlights the difference between 'improved' and what is in fact 'safe' drinking water (Onda et al., 2012) (Table 2.8).

A total of 1.8 to 3.7 billion people did not have access to safe drinking water in 2010. While the 2000 Millennium Development Goal of halving the percentage of people without access to 'improved' drinking water by 2015 was met, this is not the case for access to safe drinking water. The data is a global figure and was not broken down to the urban and rural level.

Access to safe water to some extent depends on access to safe sanitation. Without the latter, water supplies can be compromised, or exposed to 'elevated sanitary risk' as noted in Table 2.9. Onda et al. (2012) also estimates that 4.1 billion people worldwide have unsafe sanitation rather than 2.6 billion with 'unimproved' sanitation.

In 2017, JMP's data was rebased to focus on 'safe' rather than 'improved' access. 'Basic' access is broadly comparable with 'improved' access (JMP, 2017a,b) (Table 2.10). These

Table 2.8 Access to safe water worldwide in 2010.

Water supplies	Million people	Range
Low sanitary risk 'safe'	3,180	2,510–3,220
Elevated sanitary risk	1,260	740–2,130
Unsafe	1,020	746–1,610
Unknown safety	380	380
Unimproved	780	780
No data	300	300
Global Total	6,900	

Adapted from Onda et al., 2012.

Table 2.9 Percentage of people without safe drinking water.

	2000 Actual	2010 Actual	2015 Projected	2015 Target
MDG 'unimproved'	23%	12%	9%	12%
Unsafe (water quality)	37%	28%	26%	18%
Unsafe (water and sanitary risk)	53%	42%	46%	26%

Adapted from Onda et al., 2012.

Table 2.10 People without access to safe water and sanitation worldwide (billion).

Basic water	Urban	Rural	Total
2000	0.14	1.01	1.16
2015	0.20	0.68	1.10
Safe water	Urban	Rural	Total
2000	0.43	1.92	2.39
2015	0.60	1.52	2.13
Water piped to premises	Urban	Rural	Total
2000	0.43	2.21	2.64
2015	0.67	1.99	2.65
Basic sanitation	Urban	Rural	Total
2000	0.58	1.95	2.51
2015	0.67	1.69	2.35
Safe sanitation	Urban	Rural	Total
2000	1.90	2.47	4.35
2015	2.26	2.20	4.48
Household sewerage	Urban	Rural	Total
2000	1.21	2.99	4.17
2015	1.59	3.08	4.70

Adapted from JMP (2017b).

figures are in broad agreement with the estimates generated by Onda et al. (2012) with the difference that they will be broadly adopted and will refocus attention on the scale of those without access.

In urban areas, there is a link between safe access to water and sanitation and access to household piped water and sewage treatment, as demonstrated by 670 million people being classified as lacking safe sanitation although they have household sewerage.

In 2015, the United Nations unveiled its 2030 Sustainable Development Goals ('SDGs'). SDG 6 seeks to 'ensure availability and sustainable management of water and sanitation for all'. SDG 6 includes:

[6.1] Universal and equitable access to safe and affordable drinking water for all.

[6.2] Access to adequate and equitable sanitation and hygiene for all and end open defecation, paying special attention to the needs of women and girls and those in vulnerable situations.

[6.3] Improve inland water quality by reducing pollution, eliminating dumping and minimising the release of hazardous chemicals and materials, halving the proportion of untreated wastewater and substantially increasing water recycling and safe reuse globally.

To meet SDG 6 in the 140 countries highlighted by the World Bank (Hutton and Verguhese, 2016) will cost $43.1 bn pa for water and $69.4 bn pa for sanitation between 2016 and 2030, compared with capital spending current levels, estimated at $16 bn pa (Tremolet, personal communication, 2017).

Hutton and Verughese (2016) note that an additional 1,977 million urban dwellers in 140 countries will need to gain access to safe water and 3,214 million to safe sanitation by 2030 for universal access. The population of these countries the population of these countries is set to rise from 6.12 billion in 2015 to 7.14 billion by 2030. Using the current WHO criteria for access to unsafe, basic and safely managed water and sanitation, by 2030, 2.29 billion people will need water access to improve from unsafe to safe and 2.24 billion from basic to safe. For sanitation, 2.61 billion will need sanitation access to improve from unsafe/none and a further 3.16 billion will have only basic access. In these countries, 2.42 billion urban dwellers (81%) do not have their sewage treated to at least secondary level (Lloyd Owen, 2016).

2.1.6 Industrial Water Usage

Industrial consumers operate both as customers of municipal water and wastewater systems and via their own dedicated treatment systems and supplies. They typically have more flexibility to innovate with the technology and techniques they use than water utilities and often have significant financial or regulatory incentives to minimise water consumption and to prevent effluent generation.

Looking at the water usage and water intensity (the amount of water used to generate a unit of revenue), Markower (2016) notes a small overall rise in industrial water usage between 2010 and 2014 (Table 2.11) while the amount of water used to generate each $ million in GDP (Table 2.12) has fallen in the same period with the exception of where it is directly withdrawn by the company.

Table 2.11 Industrial water use, 2010 and 2014.

Million m^3 pa	2010	2014
Direct withdrawal (surface/ground)	89,067	117,220
Purchased (municipality)	9,568	9,073
Cooling water	446,982	428,773
Supply chain	658,307	694,778
Total	1,203,924	1,249,844

Adapted from Markower (2016).

Table 2.12 Value generated from industrial water.

m³ of water per $ million in revenue	2010	2014
Direct withdrawal (surface/ground)	3,700	4,200
Purchased (municipality)	400	300
Cooling water	18,400	15,300
Supply chain	27,100	24,800

Adapted from Markower (2016).

Purchasing water from a municipality is significantly more expensive than direct extraction either in manufacturing or for cooling water. Cost drives efficiency; Markower (2016) estimates that municipal water generated 14 times more value per unit than direct withdrawal and 51 times more than when used in cooling.

2.1.7 The Supply Management Paradigm

Supply management assumes that sufficient supplies can be obtained to meet increased demand in the foreseeable future. Supply management works well when supplies are plentiful and secure. As noted in Chapter 1, water, unlike telecoms and electricity is localised, as supplies are at the local or river basin level rather than national and each water source will have different characteristics. Beyond the river basin, specific supply projects have to be developed, involving further energy and infrastructure cost factors for bulk water transportation. Desalination is another option in areas near to the sea, but it carries significant costs.

Walker (2011) examined 21 projections for future water supply needs in England and Wales made between 1949 and 2009 and plotted them against the actual amount put into supply over these years. In the mid-1970s, the 10 Water Authorities and various Statutory Water Companies were supplying 13 million m^3 a day. Two projections in that decade foresaw 28 million m^3 a day being needed in 10-20 years time, two forecast 24 million m^3 and one for 20 million m^3. Supply in fact peaked at 17 million m^3 during this period and in 2010 it was 14 million m^3. If the projections from the 1970s had been adopted, the water utilities would have had to develop assets which would never in fact have been used and either continued to maintain them or have them decommissioned.

2.1.8 Funding Constraints; The Need to do More with Less Funding

The ability and the willingness to pay for water and wastewater services is one of the principal challenges facing utilities and resource managers. Full cost recovery (covering the cost of operating a service and the ability to finance the debt required to develop new assets) and sustainable cost recovery (similar to full cost recovery, blended with grants from international donors and/or the national government) ought to be the norm, but remain the exception. The only known case of a utility paying directly for all of its operating and capital expenditure costs without recourse to subsidies or debt is Copenhagen Water in Denmark.

The imbalance between current levels of capital expenditure and what is needed (GWI, 2011) is significant. GWI (2011) estimated that utilities in 2010 were spending $173 billion in capital spending compared with $384 billion pa needed to maintain services at their current level while ensuring no further overall deterioration in these assets and $534 billion pa to secure supplies and meet currently foreseen standards and demands.

Table 2.13 is the author's best estimate in 2015 of the global coverage of water and wastewater services in urban areas. In Tables 2.7 and 2.8, the lack of access to safe water was highlighted. The lack of sewerage and sewage treatment in urban areas is a contributor to an overall lack of access to safe drinking water.

The essential challenge water and sewage utilities face lies in obtaining suitable funding for maintaining, let alone extending their services. A survey commissioned by the World Bank, using data from 1999 to 2004, found that 60% of the utilities examined achieved some degree of cost recovery, especially in higher income countries. Average tariffs used were based on residential consumption of 15 m^3 of water per annum, from utilities serving 131 major cities worldwide, broken down as shown in Table 2.14.

No (or inadequate) cost recovery was interpreted as charging less than $0.20 per m^3 and partial cost recovery as charging $0.20–0.40 per m^3. Among the OECD members and in Latin America, cost recovery is the norm, but in the less developed economies in Asia and Africa it typically remains the exception.

Looking at developed economies, benchmarking data from 48 utilities in 17 countries in 2014 highlights that the financial situation has not improved in recent years, with cost recovery still remaining difficult to attain (Table 2.15).

A survey by the World Bank (Danilenko et al., 2014) collated data from 1,861 utilities serving 513 million people with water and 313 million with sewerage in 12,480 towns and cities mainly from outside the OECD. Revenues as a percentage of operating costs fell from 121% to 108% between 2000 and 2010 with tariffs charged as a percentage of household incomes also falling over this period, from 1.05% to 0.59% (Table 2.16).

Cost recovery is in inverse proportion to water usage; top performing utilities (tariffs cover at least 130% of operations and maintenance costs) had a median water consumption of 118 litres per capita per day, while the poorest performers (less than 85% cost recovery) consumed 258 litres per capita per day (Table 2.17). Much of the difference is due to leakage and unbilled water. 37% of utilities had tariffs that failed to cover their most basic O+M costs.

Table 2.13 Urban access to water and sewage services.

People in urban areas (million)	Served	Unserved	% Served
Households with piped water	3,095	862	78%
Households connected to sewage network	2,457	1,498	62%
Sewage treated to secondary level	1,342	2,613	34%
Sewage treated to tertiary level	443	3,514	11%

Adapted from WHO data (JMP, 2015) for household access to piped water and sewerage, UN DESA (2015); UN DESA (2014); OECD data, Eurostat, Global Water Markets 2016 and the author's global water infrastructure database.

Table 2.14 Water tariffs and cost recovery, 1990–2004.

	Water tariff			Utilities with of cost recovery		
	Mean ($/m³)	Median ($/m³)	Range ($/m³)	None	Partial O and M	Partial capital
Global	0.53	0.35	0.00–1.97	39%	30%	30%
By income						
High income	1.00	0.96	0.00–1.97	8%	42%	50%
Upper–medium income	0.34	0.35	0.03–0.81	39%	22%	39%
Low–medium income	0.31	0.22	0.04–0.85	37%	41%	22%
Low income	0.11	0.09	0.01–0.45	89%	9%	3%
By region or group						
OECD	1.04	1.00		6%	43%	51%
Latin America	0.41	0.39		13%	39%	48%
ME and North Africa	0.37	0.15		58%	25%	17%
East Asia and Pacific	0.25	0.20		53%	32%	16%
South and Central Asia	0.13	0.16		100%	0%	0%
Sub-Saharan Africa	0.09	0.06		100%	0%	0%

Adapted from Foster and Yepes (2005) with further data from Olivier (2007).

Table 2.15 Water costs and cost recovery in developed economies.

	Water		Wastewater	
2014 data	Median	Range	Median	Range
Cost, € per m³	1.28	0.40–2.55	–	–
Connection charge (€ pa)	–	–	180	90–350
% Of disposable household income	0.58	0.26–1.10	0.56	0.17–1.32
Cost coverage ratio	1.10	0.75–1.70	1.02	0.52–1.44

Adapted from European Benchmarking Co-operation (2015).

The fall off in sanitation coverage was especially marked in lower income countries. It is also evident that household spending on water is rising at a slower rate than operating costs and that tariffs are shrinking as a percentage of household income.

2.1.9 Affordability is a Concern, Especially in Less Equal Societies

The challenge for utilities when seeking to increase revenues is the willingness of consumers to pay more for better services. Effective average household income affordability limits of 1.4% for high income, 1.8% for medium and 2.5% for low income

Table 2.16 Utility performance in developing economies, 2000–2010.

	2000	2010
Revenues ($ per m^3)	0.34	0.81
Operations and maintenance costs ($ per m^3)	0.28	0.75
Revenue as a percentage of O+M costs	121%	108%
Revenue per person per annum ($)	18	45
Tariffs as a percentage of household income	1.05%	0.59%

Adapted from Danilenko et al. (2014).

Table 2.17 Utility coverage by country income category.

Median coverage by income category	Water (2009)	Wastewater (2010)
Low income	62%	14%
Lower-middle	81%	48%
Upper-middle	93%	77%
Upper	100%	89%

Adapted from Danilenko et al. (2014).

countries are a fair reflection as to attitudes towards affordability and willingness to pay (Tables 2.18 and 2.19).

There are a number of countries where higher tariffs than the effective limit have been broadly accepted. The scope for increases above these levels is limited. A dramatic example of willingness to pay identified is in Zambia (Klawitter, 2008), where a range of 3–10% of family income was cited as being affordable.

Smets (2008) found a wide range of tariffs in developing and transition economies, with a general understanding that in Latin America and Eastern Europe and Central Asia that the limits of affordability were identified at between 4.5% and 5.15% of household income. Specific pro-poor subsidies were noted in 13 of the countries covered in this survey, along with policy objectives to minimise the impact on poor people (keep expenses below 3–5% of household income in the UK, US, Chile and France, as well as Venezuela, Lithuania, Argentina and Indonesia). In OECD countries, average water tariffs in the late 2000s were 0.2% to 1.4% of household incomes, although for the range for

Table 2.18 Consumer perception of acceptable water tariffs.

% of household income	Range noted	Effective limit
High income (OECD)	0.5–2.5%	1.40%
Medium income	0.5–3.5%	1.80%
Low income	0.3–3.8%	2.50%

Adapted from Lloyd Owen (2009).

Table 2.19 Water tariffs as a percentage of household income.

Industrialised countries – Median	1.10%
Industrialised countries – Poor	2.60%
Developing countries – Median	2.50%
Developing countries – Poor	6.0–8.0%
Western Europe	3%
Developing countries – Without targeted support	7%
Developing countries – With target support	5%

Adapted from Smets (2008).

Table 2.20 Water tariffs as a percentage of household income in the OECD.

	Average	Bottom decile
Western Europe	0.2–1.0%	1.1–3.5%
Central and Eastern Europe	1.2–1.4%	3.9–9.0%
USA, Mexico and Canada	0.2–0.3%	1.3–3.1%
Other countries	0.3–1.0%	1.1–3.3%

Adapted from OECD (2009).

the bottom decile was 1.1% to 9.0% suggesting affordability concerns amongst the worse off (OECD, 2009) (Table 2.20).

It is evident that the ability of poorer people to pay is a challenge here. This is particularly marked in some societies. In Mexico, the average household pays 0.2% of household income for water, but this is 3.1% for the poorest decile. That creates an effective barrier when considering overall tariffs and how they affect the worst off.

2.1.10 Paying for Water and Wastewater

Sustainable water and sewage services are those that are capable of maintaining the integrity of the water cycle through an appropriate level of water abstraction and sewage treatment. As an ideal, all urban areas ought to have full household access to piped water and mains sanitation, all urban sewage to be collected and treated to at least secondary level and all urban areas to have storm sewerage. Table 2.21 is based upon the author's estimates for the cost of capital spending to achieve this in urban areas worldwide by 2050 (2010 prices).

This is an ideal, albeit a challenging one. It needs to be reconciled with the ability of people to pay for new infrastructure. A series of medium-term estimates for urban water and sewerage spending needs in 69 countries for 2010–2029 was developed by the author (Lloyd Owen, 2009) and applied them to the effective affordability limits outlined above. The 'medium' capital spending scenario was based on meeting various national and international targets for service coverage and infrastructure maintenance. In every case, forecast tariff revenues were less than the forecast spending needs, as outlined in Table 2.22.

Table 2.21 Estimated capital costs for developing sustainable
urban water and wastewater services by 2050.

$ billion, 2015–50	
Water distribution	1,421
Water treatment	1,112
Sewerage	2,259
Sewage treatment	2,041
Storm sewerage	954
Metering	169
Global total	8,226

Developed from the author's water infrastructure database. An
earlier version was published as Lloyd Owen (2011).

Table 2.22 Forecast excess of spending over revenues for three capital spending scenarios, 2010–29.

Revenue-spending gap ($ billion)	Capital spending scenario		
	Low	Medium	High
North America	−344.7	−435.1	−702.3
Western Europe	−92.0	−231.2	−344.0
Rest of Europe	−15.8	−35.0	−54.2
Developed Asia	−352.6	−419.0	−481.5
Latin America	−13.3	−38.3	−55.3
MENA	−110.4	−157.1	−189.8
Sub-Saharan Africa	−0.5	−3.6	−11.0
E and SE Asia	−39.5	−146.6	−282.0
South Asia	−80.2	−120.2	−197.6
Total	−1,049.0	−1,522.6	−2,296.9

Adapted from Lloyd Owen (2009).

Overall, the low spending forecast pointed towards a shortfall of $52 billion per
annum, the medium forecast to $76 billion pa and the high forecast to a $115 billion pa
shortfall. As with the GWI (2011) data outlined in 2.1.8 above, there is a pronounced
gap between what is being spent, what needs to be spent and the revenues needed to
finance this spending.

2.2 The Impact of Climate Change

Climate change brings in a new area of uncertainty with forecasts ranging between two
and seven billion people facing absolute water scarcity by 2075, according to a series of
climate scenarios (Falkenmark, 2007).

Willett (2007) notes that between 1973 and 1999 a significant global increase of humidity in the atmosphere was primarily due to human activity. Such a trend is set to influence where rain falls and how heavily it falls. Warmer rivers and streams hold less dissolved oxygen and are more vulnerable to nutrient loadings. In the uplands of Northern England, the past two decades have seen a shift towards heavy winter rainfall and an 'almost complete absence of heavy summer rainfall', which was 'in marked contrast to the patterns seen in lowland areas' (Burt and Ferranti, 2011). Globally, daily rainfall extremes have risen by 1–2% a decade since the 1950s (Donat et al., 2016) reflecting the increased amount of water vapour in the atmosphere because of rising temperatures.

Projections point to more rainfall and river run-off in high latitudes and the tropics and less rainfall and river run-off in sub-tropical and other regions, increasing the dry areas. Globally, the IPCC concludes that by 2050 twice the land area will be subject to reduced precipitation than increased precipitation as a consequence of climate change. The shift towards more varied and extreme rainfall seen over the last century will increase and rising water temperatures will continue. Forecasts for high latitudes water stress are 'very likely' to be reduced by climate change (IPCC, 2008) (Table 2.23).

Increasing temperatures intensify water demand, especially for irrigation agriculture and through human activities such as watering gardens and using swimming pools, and the need for cooling water for electric power and industrial plants. Changing seasonal patterns of precipitation also modifies demands for irrigation, particularly in regions with soils of low water-storage capacity.

Between 1994 and 2006, there has been an 18% rise in water discharge into oceans from rivers and glaciers, or a 1.5% per annum rise in run-off. This may be due to higher evaporation from the oceans, increasing the intensity of the water cycle (Syed et al., 2010). There Indications of higher temperatures above the oceans mean faster evaporation (and more rain in general, but not necessarily on land) and more storms.

Stern (2007) considers the forecast impact of climate change between 2050 and 2085 in terms of what happens at each degree Celsius rise (Table 2.24). Putting these into context, the Paris Agreement of 2015 seeks to limit the rise to two degrees Celsius by 2050 (UN FCCC, 2015).

Table 2.23 Predicted impacts of climate change on water management.

- Exacerbation of pressures on water resources and their management;
- Higher frequency of extreme drought events;
- Inland waters will be warmer and therefore less able to absorb oxygen and less able to tolerate nutrient build-ups;
- Inland waters more vulnerable to over-abstraction and pollutant build-up;
- Increase in forestry productivity and agricultural productivity in certain areas, with increased water demand as a result;
- Decreased food security in Africa and Asia;
- Modification in patterns of tourism and outdoor recreational activities and changes in water demand where recreation increases or decreases;
- Increased need for storm sewerage systems and separate storm and foul water systems to deal with extreme rainfall.

Adapted from IPCC (2008).

Table 2.24 The impact of climate change by 2050–85.

1°C	50 million affected by loss of Andean glaciers
2°C	20–30% decrease in rainfall in Southern Africa and the Mediterranean
	10–20% increase in rainfall in Northern Europe and South Asia
3°C	Southern Europe has serious droughts every 10 years
	1–4 billion people exposed to water shortages (ME and Africa)
	1–5 billion people face greater flood risk (S and E Asia)
4°C	30–50% decrease in rainfall in Southern Africa and the Mediterranean
5°C	750 million affected by loss of Himalayan glaciers

Adapted from Stern (2007).

In the UK, the core range of forecasts on the low emissions scenario, summer rainfall is forecast to fall by 1–15% by 2025 and winter rainfall to rise by 3–13%, with somewhat larger changes for the higher emissions scenarios (DEFRA, 2009).

2.2.1 The Cost of Adapting to a Changing Climate

The United Nations Framework Convention on Climate Change made a global estimate of $425–531 billion needing to be spent in the 23 years between 2008 and 2030, or $18–23 billion pa (Bates et al., 2008). Most of the extra spending is set to be needed on reservoirs, desalination, water recovery and reuse and irrigation efficiency in Africa and Asia: Africa ($131–138 billion), Asia ($238–288 billion), South America ($12–20 billion), Europe and North America ($37–86 billion) and Australia and New Zealand ($1–34 billion). An assessment concentrating of the direct consequences of climate change points to $11 billion pa being needed to 2030 (Kirschen, 2007), $2 billion pa in the developed world and $9 billion pa in the developing world.

2.3 Leakage and Water Losses

Unaccounted-for water, the difference between water put into the mains network and what reaches the customer is an increasing concern. It represents a waste of potable water and water that could have been earning revenues for the utility. The potential contribution through reduced leakage by halving water losses in low and middle income cities would provide enough water to satisfy a further 130 million people and improve utility cash-flow by $4 billion per annum (Liemberger, 2008). In the late 1990s, there were 42% leakage rates in major cities in Africa and Latin America and 39% losses in Asia (WHO/UNICEF, 2002). A survey by the Asian Development Bank concluded that Asia loses 28.7 billion m^3 of water each year through leakage with a minimum value of this loss of $8.6 billion pa (Frauendorfer and Liemburger, 2010).

The American Water Works Association recommends that 10% unaccounted-for water is a benchmark for the well-managed utility, with action needed when it goes above 25%. Distribution losses of 6% to 14% occur amongst utilities seen as good performers in England, the Netherlands and Germany.

Distribution losses also matter more where water scarcity is more pressing. At 10.8%, leakage at Sydney Water in 2004 was too high given the context of supplies. By 2010, leakage was down to 6.6%, a total saving of 66 million litres a day since 1999 (Sydney Water, 2010). In this case, 21% of the leakage reduction came from pressure management rather than repairing pipes. In Riyadh, 1.1 million m^3 of water per day was lost in 2005, 31% of the supply and the equivalent of nine otherwise unneeded desalination plants (Al-Musallam, 2007). By spending $400 million in basic network repairs in Riyadh, the city believes it can save $2.1 billion over 20 years in avoiding the need for new desalination plants.

2.4 Water Efficiency and Demand Management

In order for future water supplies to meet anticipated demand, we need to consider how that demand can be managed, preferably within available supplies. As noted in 2.1.7 above (Walker, 2011), under a pure supply management approach during the 1970s, water storage and supply assets in England and Wales would have been developed to supply two to three times more water than was in fact needed.

2.4.1 Demand Management and Consumer Behaviour

Demand management can be defined as a means of enabling consumers to appreciate the direct and indirect cost of the water they consume, how their consumption can be changed and by how much it can be changed.

In order for consumer behaviour to be modified, appropriate motivations are needed. Various methods are being used for domestic customers, including labelling schemes for buying water-efficient domestic goods and through water metering. Commercial customers can be encouraged to recycle their water. For example, hotels and casinos in Las Vegas accounted for 3% of the city's water consumption in 2010 after a series of conservation measures. As a result, water outflow from Lake Mead fell by 15% between 1980 and 2010 when the population of Las Vegas and the state of Nevada rose by 254% and 238% respectively (data from lakemead.water.data.com). Industrial customers can be motivated to internalise their water usage and mechanisms exist for improving agricultural and recreational water efficiency. All such behaviour changes usually require regulatory or economic incentives.

To date, demand management has mainly been achieved through traditional economic and technological approaches. The interests and expectations of customers have not usually been taken into much consideration by utilities. The one-way flow of information from manual meter readings or property-based assessments for water tariffs limits the scope for influencing consumer behaviour. It fails to enable utilities to appreciate the variable nature of consumer behaviour and in turn, utilities typically communicate with their customers in a simplistic manner, which gives little room to cater for individual needs or preferences.

2.4.2 Balancing Water Use; Seasonal Demand and Availability

Patterns of demand vary over differing time scales, within the day and across weeks and years. By smoothing out demand for water, the scale and scope of redundant assets is eased, for example, by encouraging the use of domestic appliances at low demand times and incentivising customers to conserve water during drier seasons. Daily tariffs have been used for electricity supplies for some time, for example in the UK the 'Economy 7' tariff was introduced in 1978 (Electricity Council, 1987).

Smoothing out patterns and perturbations in water demand enables assets to be used in the most effective manner. Instead of assets being geared towards managing occasional peaks of demand, they can be better used when their capacity better reflects overall demand.

This would also give more headroom for dealing with genuinely exceptional peaks in demand either due to unexpectedly high need or extreme weather events. Such extremes can also be smoothed out through specific demand management measures. For example, smart flood analysis of a catchment area offers the potential to deploy natural and built assets to delay some of the discharge of flood waters into river systems by holding the water upstream so that it is released over a longer period. Another example would be in the integrated monitoring of management of storm and foul sewerage systems to provide greater capacity for exceptional events as well as ways of modifying customer behaviour to minimise discharges during such peak periods.

2.4.3 Water Efficiency – The Demands of Demand Management

Household goods have an impact on water demand, even without the need to alter consumer behaviour. Table 2.25 compares traditional household goods in Europe to ones designed with water efficiency in mind (Dworak, 2007).

Some of these reductions come from providing less water (a slimmer bath or a shower with less power) while others use water more appropriately (a dual flush versus a single flush loo) or efficiently (washing machines and dishwashers).

A reasonable domestic water consumption target in Western Europe would be 110–130 l/cap/day as 120–128 l/cap/day is already the norm in parts of Germany and the Netherlands (Green, 2010), while it is 100 l/c/d in Copenhagen after a series of demand management measures (Chapter 5.8.1). There are some unintended consequences in lowering water consumption. Longer residence times of drinking water within German distribution networks had meant that water treatment standards have needed to be

Table 2.25 Water consumption by standard and water efficient household goods.

(Litres per household/day)	Standard	Efficient	Reduction
Lavatory	57–87	39	32–52%
Shower	45–54	43	3–44%
Bath	71	53	26%
Taps	10	8.5	15%
Washing machine	26	17.4–19.6	25–33%
Dishwasher	8.7	5.2–6.1	30–40%
Total	237–280	167–169	29–41%

Adapted from Dworak (2007).

enhanced to prevent bacterial contamination building up and low flows affect the flushing of sewage into the sewer mains.

2.4.4   Water Metering

In some countries, urban water metering has long been accepted, for example in France, Chile and Singapore. In others, such as the UK and Norway it is either under gradual adoption or is still not being actively considered. For the former, smart metering is about transforming the possibilities of an accepted approach. For the latter, it is about providing a wholly new customer experience. Each poses cultural and behavioural challenges for utilities and their customers.

2.4.4.1   The Development of Metering in England and Wales

With the exception of the 50,000 households in the Isle of Wight (served by Southern Water) which meters installed for trials in 1989, there was effectively no domestic metering in England and Wales prior to the privatisation of the Water and Sewage Companies in 1989. Using company data collected by Water UK for 2014–15 (Water UK, 2015), along with company data collected in the Ofwat 2000, 2005 and 2010 June Returns (Ofwat, 2000, 2005 and 2010), the installation rate of new meters and metering coverage can be traced for the past 25 years (Table 2.26).

During this period, household sizes has decreased and the population has increased. For example, during 2010–15, there were 2.61 million fewer people with unmetered supplies against 6.80 more with metered supplies, a net increase of 4.19 million in five years.

Given that people who opt to have water meters will tend to be those who stand to benefit the most, it would be expected that the difference between metered and unmetered consumption would decrease as meter penetration increases. In fact, the opposite took place as seen in Table 2.27, the reduction in water consumption was 9-10% in 2000 and 2005 and 18% in 2010 and 2015. While unmetered use does not appear to have changed much during this period, there is a fairly consistent decline in metered consumption. Given that more of the higher usage customers are being metered, this would suggest that consumer behaviour is being modified by metering.

In the UK customers with a meter pay £100 on average less than customers without a meter, with a difference of £400 in the South West Water region, where bills are the highest. Traditional one-way meters have achieved a 12% reduction in water consumed and also a 12% reduction in electricity consumption by the utilities, equivalent to 1,050 GWh (Savic, 2015).

Table 2.26 Development of water metering installation in England and Wales, 1990–2015.

	Meters pa	Coverage
1990–00	715,000	14%
2000–05	904,000	22%
2005–10	1,151,000	33%
2010–15	1,479,000	43%

Adapted from Ofwat (2000, 2005, 2010) and Water UK (2015).

Table 2.27 Household water usage in England and Wales.

L/cap/day	Unmetered	Metered
1999–00	149	135
2004–05	152	138
2009–10	157	128
2014–15	151	123

Adapted from Ofwat, (2000, 2005, 2010) and Water UK (2015).

A study by Arqiva, Artesia and Sensus in the UK to quantify how technology selection might impact water companies, consumers, and the environment compared drive by (one-way communications) and fully smart (two-way communications) domestic water metering (Table 2.28).

Water metering is usually seen as being concerned with altering consumer water demand. Section 2.5 below considers the impact that water consumption has on energy. It is evident that water metering and modifying customer behaviour can also have a significant impact on domestic energy usage.

2.5 Lowering Energy Usage

Water is energy intensive. In an extreme case, Jordan is spending $5.4 billion pa on pumping water from reservoirs to its major population centres, equivalent to 15% of its GDP (Hirtenstein, 2016). In California, 4% of energy used in the state is accounted for by water supply and treatment and a further 14% by end users of water (Klein, 2005). The water sector contributes directly 1% of the UK's greenhouse gases (Defra, 2008) and 2–4% indirectly when taking into account water heating and transporting. Direct and indirect use accounts for 5% of energy consumption in the USA (Rothausen and Conway, 2011).

Table 2.28 One-way and two-way metering compared.

Benefits	AMR drive by metering	Smart metering
	One-way	Two-way
Gross benefits	£0.3 billion	£4.4 billion
Consumer benefits - bill savings	£11/household pa	£40/household pa
Water savings	40 million m^3 pa	294 million m^3 pa
Carbon reduction	8 million tonnes pa	31 million tonnes pa

Adapted from Slater (2014).

2.5.1 The Cost of Energy

Energy is expensive and so it makes good sense to minimise its use. In England and Wales, 18% of water and sewage utility operating spending is accounted for by power costs in 2014–15, compared with 11% in 2004–05 (Table 2.29).

The low energy cost for sludge treatment reflects the emergence of sludge to energy applications. Indeed, two of the Water and Sewerage Companies (Northumbrian Water and Severn Trent) generated energy revenues rather than costs.

Areas where scope for efficiency exist include water and sewage pumping, the sewage treatment process and sludge to energy. The less energy consumed, the greater the sector's contribution in meeting national and international carbon reduction targets.

2.5.2 Where Energy is Consumed

Direct energy use by the water sector relates to energy being used to pump water to a utility (bulk transfer) from a remote location (a river, reservoir or lake outside the catchment area) or from groundwater sources. The water itself may have been desalinated sea or brackish water, where energy is needed to either distil the water or to drive it across a membrane. Water reuse for indirect or direct potable applications is also membrane based, although less energy intensive.

Within the utility network, water may need to pass through filtration units as part of the treatment process and to be pumped into the distribution network to its customers. The water needs to be delivered at an agreed pressure range. Additional pumping may be needed in multi-storey buildings.

These processes are repeated in sewage collection and treatment, with a particular emphasis on the energy needed when handling the sewage and sludge during treatment processes and preparing the sludge for disposal.

Table 2.29 Power costs and total operating costs in England and Wales, 2014–15.

YE 31/03/2015 (£ million)	Power cost	Operating cost	Power as %
Water resources	51.4	265.3	19%
Raw water distribution	30.6	89.0	34%
Water treatment	138.1	575.6	24%
Treated water distribution	88.1	890.6	10%
Total – water	308.2	1,820.5	17%
Sewerage	64.4	378.3	17%
Sewage treatment	196.4	789.7	25%
Sludge treatment	31.9	307.8	10%
Sludge disposal	0.2	78.9	0%
Total – sewage	292.9	1,554.7	19%
Total – water and sewage	601.1	3,375.2	18%

Adapted from Water UK (2015).

Indirect energy use covers energy consumed at the point of delivery. The greatest impact comes from heating water for baths, showers, washing machines and preparing food and beverages. Energy is also needed to drive water based applications such as power showers, and dish and clothes washing machines.

The UK used 8.8 TWh of electricity in the urban water cycle in 2011 and a further 81 TWh to provide household hot water. This is in fact falling, with 125 TWh used in 1970, while water heating as a percentage of household energy has fallen from 28% in 1970 to 18% in 2011 (Savic, 2015).

Research by the Energy Savings Trust (McCombie, 2014, 2015) looked at 86,000 household water and energy accounts in the UK. The average energy bill in the survey was for £1,320 against £385 for water, yet 8% of the people surveyed appreciate the role of water usage in their energy bills. So, there is room to address customer behaviour. The survey found that 54% of household water is heated for various uses, with 68% being used in the bathroom (Table 2.30).

The average shower lasts for 7–8 minutes, with 87% taking less than 10 minutes (45% 1–5 minutes and 42% 6–10 minutes) against 13% taking longer (12% 10–20 minutes and 1% 20+ minutes). 16% of households have water-efficient shower units (5% eco-power and 11% eco-mixer), with 49% using water intensive units (31% standard mixer and 18% standard power) and using 35% electric mixers. 41% of lavatory units were found to be dual flush, 6% in homes that were built in 1940–80 against 57% in those built since 2001.

It is intriguing to observe the disconnected nature of water and energy consumption. Water is used to cool the plant during power generation. Energy is in turn used to heat water for consumers.

2.5.3 Energy Efficiency

Treatment and distribution processes can be made more efficient. There is a considerable scope for lowering the energy intensity of pumping; by improving the efficiency of the pumps themselves, by optimising their deployment and my minimising the amount of pumping that needs to be carried out.

Pumps in general account for approximately 10% of global energy consumption. Pumps are typically inefficient (newer designs doing more for less energy), over-deployed (excess capacity) and not utilised in the most effective manner (location, integration and timing). Indirectly, pump management also affects distribution system

Table 2.30 Household water use in the UK, by appliance.

Cold water	Hot water
Garden 1%	Shower 25%
Car 1%	Bath 8%
Loo 22%	Dishwasher 1%
Cold taps 22%	Hand wash dishes 4%
	Washing machine 9%
	Bathroom hot tap 7%

Adapted from McCombie (2014).

performance as unnecessarily high pressure in the water mains drives up distribution losses. If network pressure can be effectively managed, energy is saved both from less pumping and less water being needed.

As discussed above, avoiding the excessive use of domestic water and heated water in particular has a significant impact on water-related energy consumption.

2.5.4 Turning Wastewater into a Resource

Sludge to energy is of particular importance because of the greenhouse gas impact of methane generated by sewage. Optimising the conversion of sludge to methane which is then used to generate electricity means that a wastewater treatment plant's energy needs can be internally sourced, impacting the processes' carbon footprint through lower emissions and electricity consumption.

Wastewater has an internal chemical energy of 7.6 kilojoules per litre, and this does not take into account the energy stored in industrial and commercial wastewater (Heidrich, Curtis and Dolfing, 2011). Power generated from sludge reached 2,939 GWh in 2010 worldwide, with 35% of this coming from Germany (GlobalData, 2011). In theory, wastewater contains seven to ten times the energy it takes to treat it. In Germany, trails found that co-fermenting municipal wastewater generated enough energy at the on-site gas engines to cover 113% of the energy the treatment facility used (Schwarzenbeck, Pfeiffer and Bomball, 2008). The Egå Renseanlæg wastewater treatment plant in Aarhus Denmark, anticipates generating 150% of its energy needs when it enters service in July 2016 (Freyburg, 2016).

Sludge to energy, when combined with measures to minimise energy use in the rest of the water and sewerage network have the potential to significantly reduce a utility's carbon footprint as well as lowering its operating costs.

2.6   Appreciating Asset Condition and its Effective Performance

As noted in section 2.1, finance is a particular challenge for the water sector and this is being exacerbated by the challenges posed by demographic change (section 2.1) and climate change (section 2.2) as well as assets that are not performing efficiently (section 2.3).

Table 2.31 highlights the asset intensive nature of water utilities. In England and Wales, the fully privatised water utilities are financially self-sufficient; they have to use tariffs to finance their operations and to finance any further fund-raising needed. Such are the estimated Gross Replacement Costs of their assets that 83 years of their combined cash flow would be needed to replace these assets, irrespective of any further spending and at current levels; 110 years of capital spending would be needed to replace these assets. There is an evident need to ensure that assets are effectively used, their operating lives maximised and that any new assets are genuinely needed.

These figures are for the regulated water and sewerage activities only. Capital spending here is investment in fixed assets and infrastructure renewals charge. The gross replacement costs (GRC) of their assets includes infrastructure and non-infrastructure assets and is net of depreciation. The regulated asset value (RAV) accounts for assets

Table 2.31 Operating statistics for the English and Welsh water utilities, 2014–15.

FY 31/03/2015 (£ million)	Water	Sewage	Combined
Gross replacement cost (GRC)	161,484	429,548	591,032
Regulated asset value (RAV)	25,710	37,240	64,750
Revenues	5,801	6,026	11,885
Operating profit	1,564	1,923	3,497
Revenues as % of GRC	3.6%	1.4%	2.0%
Operating profit as % of GRC	1.0%	0.5%	0.6%
Revenues as % of RAV	21.1%	16.2%	18.4%
Operating profit as % of RAV	5.7%	5.2%	5.4%
Cash flow as % of GRC	–	–	1.2%
Capital spending as % GRC	–	–	0.9%
Cash flow as % NAV	–	–	11.2%
Capital spending as % NAV	–	–	8.0%

Adapted from Water UK (2015).

added since privatisation in 1989 which are used by Ofwat, the sector's regulator, for assessing allowable returns on investment by the various companies.

Many of these assets are also hard to reach, especially the underground systems. The treated water distribution (mainly the water mains) assets have a GRC of £113.8 billion, or 29% of the water GRC and the sewage collection networks have a GRC of £377.4 billion or 89% of the sewage GRC. Finally, in GRC terms, water assets are worth £1,743 per person (57.6 million people served) and sewage assets are worth £8,607 per person (56.9 million people served).

2.6.1 Improvements in Asset Efficiency and Operating Costs

Utilities need to develop the capability to micro manage their assets to the best effect. It has been noted in section 2.4 that demand patterns can be smoothed to reduce overall capacity needs. In 2.5 the potential to lower energy costs and to lower distribution losses was noted. With a proactive cycle of maintenance, refurbishment (which may include updating and expanding) and replacement based upon the actual condition of the assets and how this is affecting their performances, utilities can focus on what needs to be done rather than what they assume requires doing.

2.6.2 The Need to Understand Underground Assets

In urban areas with formal water and sewerage services, most utility assets are buried underground. This is especially the case in centralised utility systems, where single water and sewage treatment plants serve a large number of customers, from tens of thousands right up to millions of people. There is a need to be able to assess underground assets in an efficient and non-invasive manner. These assets have traditionally been managed on assumed operating lifetimes rather than an appreciation of their actual operability.

Traditional pipeline management and maintenance offers a general appreciation where leaking or otherwise damaged parts of the network are. In addition, the work required in digging up stretches of the network work against medium-term pipeline rehabilitation in favour of comprehensive replacement.

Pipelines degrade at different rates owning both to external (soil conditions and ground profiles) and internal factors (the nature of the water/wastewater flowing through and how they are managed). When these assets can be assessed in a timely and non-invasive manner, both in terms of inspecting the walls and considering leakage under different conditions, there is more room for appropriate maintenance and repair and rehabilitation programmes to maximise their effective operating lives. A fully developed pipeline management strategy needs to be able to account for these variables.

2.6.3 Pumps and Potential Savings

The pump manufacturer Grundfos estimates that pumping in all applications account for 10% of global energy consumption, yet 90% of pumps in use are not optimised. Pump optimisation could reduce global energy consumption by 4% (Riis, 2015). There are practical reasons for seeking to improve pump efficiency. 5% of a pump's life cycle costs go on purchasing the unit, against 85% on power consumption and 10% on service and maintenance. Energy use can be reduced by 20% through buying a more modern unit. More than 20% can be saved through using pumps in a more efficient manner. Pumping accounts for 89% of power consumption in the water distribution network (Bunn, 2015); Office, heating, ventilation, air conditioning, and lighting (6%), backwash (5%), compared with high zone pumps (21%), main zone pumps (37%) and raw water pumps (31%).

2.6.4 The Scope for Savings

An analysis of UK water utilities and their customers identified £3,216 million in potential operating expenditure savings that could be brought about through innovative practices such as smart water (Table 2.32).

Evan if 20–50% of these savings were realisable in practice, this would have a significant impact on driving down costs and improving revenues.

Table 2.32 Potential efficiency savings for UK water utilities and users.

Operational benefits	
Improved demand forecasting	£64 million (2%)
Network monitoring benefits	£87 million (3%)
Reduced 3rd party liabilities	£318 million (10%)
Recovered revenue (faulty meters)	£408 million (13%)
Reduced energy consumption	£1,047 million (32%)
Leakage repair efficiency	£693 million (21%)
Customer service savings	£599 million (19%)

Adapted from Slater (2014).

Conclusions

Water utilities and other water consumers face a large and often interlinked set of challenges. What they have in common is the asset intensity of these services in relation to the revenues these services generate. The exceptions (such as agriculture) are where water is a fundamentally under-priced resource are facing resource conflicts which are set to alter the way this resource is valued.

Outside Europe, population growth will be a major driver of demand until at least 2050–80, especially in Sub-Saharan Africa. This, combined with urbanisation is causing a dramatic growth in the number and size of major cities worldwide. The rate of urbanisation is set to be greatest in Sub-Saharan Africa while the largest scale of urbanisation is taking place in South and South East Asia. Urbanisation brings economic development, which allied with international initiatives to provide universal household access to safe drinking water are in turn driving up per capita demand.

Larger cities, with a higher per capita water demand are placing new pressures on the capacity of catchment areas to adequately supply them. This means that the traditional supply management paradigm is neither sustainable nor feasible. With the exception of water reuse, alternative water resources have a limited capacity to cater for this demand, especially when considering their cost and energy intensity.

Water utilities are expected to guarantee the security and quality of their supplies. To date, this has usually been carried out by responding to events. When a failure occurs, it is noted and in turn it is addressed. Through the integration of communications, data processing and data capture, utilities can move from being reactive to proactive and either anticipating events or reacting to them quickly, in a manner that builds customer confidence in their services. The faster data can be obtained, transmitted and interpreted, the more those events can be controlled. More rapid and accurate monitoring allows utility operators to have greater confidence in for example drinking water quality (public health) and river water quality (environmental compliance).

By avoiding the development of excess assets and capacity and optimising the utility and use of extant assets, utilities can lower their capital and operating costs at a time when there appears to be little prospect of funding flows increasing to the point whereby systems can be upgraded and expanded using tariffs alone. This also increases the scope for a utility to address affordability concerns, especially amongst their poorer customers.

References

Al-Musallam L B A (2007) Urban Water Sector Restructuring in Saudi Arabia. Presentation to the GWI Conference 2–3rd April 2007, Barcelona, Spain.

Bates B C, Kundzewicz Z W, Wu S and Palutikof J P, eds. (2008) Climate Change and Water. Technical Paper of the Intergovernmental Panel on Climate Change, IPCC Secretariat, Geneva, Switzerland.

Bunn S (2015) What is the energy savings potential in a water distribution system? A closer look at pumps. Presentation at the SWAN Forum 2015 Smart Water: The time is now! London, 29–30th April 2015.

Burt T P and Ferranti E J S (2011) Changing patterns of heavy rainfall in upland areas: a case study from northern England. International Journal of Climatology 32: 518–532, doi: 10.1002/joc.2287.

Danilenko A, van den Berg C, Macheve B and Moffitt L J (2014) The IBNET Water Supply and Sanitation Blue Book 2014: The International Benchmarking Network for Water and Sanitation Utilities Databook. World Bank, Washington DC.

Defra (2009) UK Climate Change Projections 2009 – planning for our future climate. Defra, London, UK.

Defra (2008) Future Water: The Government's water strategy for England. Defra, London, UK.

Donat, M G, Lowry A L, Alexander L V, O'Gorman P A and Maher N (2016) More extreme precipitation in the world's dry and wet regions. Nature Climate Change, doi: 10.1038/nclimate2941 7th March 2016.

Dworak T, et al. (2007) EU Water Saving Potential. EU Environmental Protection Directorate, Berlin, Germany.

EA/NRW (2013) Water stressed areas, final classification. Environment Agency/Natural Resources Wales, Bristol, UK.

Electricity Council (1987) Electricity Supply in the United Kingdom: A Chronology – From the beginnings of the industry to 31 December 1985. 4th edn. The Electricity Council, London.

European Benchmarking Co-operation (2015) Learning from International Best Practices, 2014. EBC Foundation, The Hague, Netherlands.

Eurostat (http://ec.europa.eu/eurostat/web/environment/water/database).

Falkenmark M and Lindh G (1976) Water for a starving world. Westview Press, Boulder, Co, USA.

Falkenmark M and Lindh G (1993) Water and economic development. In: Gleick P H, ed. Water in Crisis: pp. 80–91. Oxford University Press, New York, USA.

Falkenmark M, Berntell A, Jagerskog A, Lundqvist J, Matz M and Tropp H (2007) On the Verge of a New Water Scarcity: A Call for Good Governance and Human Ingenuity. SIWI Policy Brief. SIWI, Stockholm, Sweden.

FAO (2010) Towards 2030/2050. UN FAO, Rome, Italy.

Foster V and Yepes T (2005) Is cost recovery a feasible objective for water and electricity? A background paper commissioned for the World Bank, and quoted in Fay, M and Morrison M (2005) Infrastructure in Latin America and the Caribbean: Recent Developments and Key Challenges. Volume I: Main Report and Volume II: Annexes, Report No. 32640-LCR, World Bank, Washington DC, USA.

Frauendorfer R and Liemburger R (2010) The Issues and Challenges of Reducing Non-Revenue Water. ADB, Manila, Philippines.

Freyberg T (2016) Denmark kick-starts energy-positive wastewater treatment project. WWi magazine, February 5, 2016.

Gleick P H, ed. (1993) Water in Crisis: A Guide to the World's Fresh Water Resources. Oxford University Press, New York, USA.

GlobalData (2011) Power Generation from Wastewater Treatment Sludge – Power Generation, Sludge Management, Regulations and Key Country Analysis to 2020.

Global Water Intelligence (2011) Global Water Markets 2011: Meeting the World's Water and Wastewater Needs until 2016. Media Analytics Limited, Oxford.

Green C (2010) Presentation to CIWEM Surface Management Conference, SOAS, London, UK, 2nd June 2010.

GWM 2016 (2015) Global Water Markets 2016: Media Analytics Limited, Oxford.

Heidrich E S, Curtis T P and Dolfing J (2011) Determination of the internal chemical energy of wastewater. Environmental Science and Technology 2011 Jan 15; 45(2): 827–32.

Hirtenstein A (2016) Refugee Influx Pushes Government of Jordan to Invest in Solar. Bloomberg Business, February 18, 2016.

Howard G and Bartram J (2003) Domestic Water Quantity, Service Level and Health, Page 22. World Health Organization, Geneva.

Hutton G and Varughese M (2016) The Costs of Meeting the 2030 Sustainable Development Goal Targets on Drinking Water, Sanitation, and Hygiene. WSP/World Bank Technical Paper 103171.

IPCC (2008) Climate change and water: IPCC Technical Paper IV. IPCC Secretariat, Geneva, Switzerland.

JMP (2015) Progress on Sanitation and Drinking Water: 2015 Update and MDG Assessment, JMP UNICEF/WHO, Geneva, Switzerland.

JMP (2017a) Progress on Drinking Water, Sanitation and Hygiene: 2015 Updated and SDG Baselines, Main Report, JMP UNICEF/WHO, Geneva, Switzerland.

JMP (2017b) Progress on Drinking Water, Sanitation and Hygiene: 2015 Updated and SDG Baselines, Annexes JMP UNICEF/WHO, Geneva, Switzerland.

Kirshen P (2007) Adaptation Options and Costs in Water Supply. A report to the UNFCCC Financial and Technical Support Division. http://unfccc.int/cooperation_and_support/financial_mechanism/financial_mechanism_gef/items/4054.php.

Klawitter S (2008) Full Cost Recovery, Affordability, Subsidies and the Poor – Zambian Experience, presentation to the International Conference on the Right to Water and Sanitation in Theory and Practice, Oslo Norway.

Klein G (2005) California's Water–Energy Relationship. California Energy Commission. CEC-700-2005-011-SF. San Francisco, California, USA.

Kummu M, Ward P J, de Moel H and Varis O (2010) Is physical water scarcity a new phenomenon? Global assessment of water shortage over the last two millennia. Environmental Research Letters 5: 034006.

Liemberger R (2008) The non-revenue water challenge in low and middle income countries. Water 21, June 2008, pp. 48–50.

Lloyd Owen D A (2009) Tapping Liquidity: Financing Water and Wastewater 2010-2029. Thomson Reuters, London, UK.

Lloyd Owen D A (2011) Infrastructure needs for the water sector. OECD, Paris.

Lloyd Owen D A (2016) InDepth: The Arup Water Yearbook 2015–16, Arup, London, UK.

Markower J (2016) State of green business 2016. GreenBiz Group Inc.

McCombie D (2014) At home with water. Presentation at the SMi, Smart Water Systems Conference, London, 28–29th April 2014.

McCombie, D (2015) Smart water and regulation. Presentation at the SWAN Forum 2015 Smart Water: The time is now! London, 29–30th April 2015.

Molden D, ed. (2007) Water for Food, Water for Life: A Comprehensive Assessment of Water Management in Agriculture. Earthscan, London, UK and International Water Management Institute, Colombo, Sri Lanka.

Muhairwe W T (2009) Fostering Improved Performance through Internal Contractualisation. Presentation to the 5th World Water Forum, Istanbul, Turkey, March 2009.

Neave G (2009) Advanced Anaerobic Digestion: More Gas from Sewage Sludge. Renewable Energy World, April 2009.

OECD (2008) OECD Environmental Outlook to 2030. OECD, Paris, France.

OECD (2009) Managing water for all: An OECD perspective on pricing and financing, OECD, Paris.

OECD (2016) Waste water treatment (indicator). doi: 10.1787/ef27a39d-en.

Olivier A (2007) Affordability: Principles and practice, presentation to Pricing water services: economic efficiency, revenue efficiency and affordability, OECD Expert Meeting, 14 November 2007, Paris, France.

Onda K, LoBuglio J and Bartram J (2012) Global access to safe water: Accounting for water quality and the resulting impact on MDG progress. International Journal of Environmental Research and Public Health 9: 880–94.

Postel S L, Gretchen C D and Ehrlich P R (1996) Human appropriation of renewable fresh water. Science 271: 785–88, 9th February 1996.

Riis M (2015) Energy savings potential in a water distribution system. Presentation at the SWAN Forum 2015 Smart Water: The time is now! London, 29–30th April 2015.

Rothausen, S G S A and Conway D (2011) Greenhouse-gas emissions from energy use in the water sector. Nature Climate Change 1: 210–19.

Savic D (2015) Metering and the Hidden Water-Energy Nexus. Presentation at the SWAN Forum 2015 Smart Water: The time is now! London, 29–30th April 2015.

Schwarzenbeck N, Pfeiffer W and Bomball E (2008) Can a wastewater treatment plant be a power plant? A case study. Water Science Technology 57 (10: 1555–61.

Slater A (2014) Smart Water Systems – Using the Network. Presentation at the SMi, Smart Water Systems Conference, London, 28–29th April 2014.

Smets H (2008) Water for domestic uses at an affordable price, presentation to the International Conference on the Right to Water and Sanitation in Theory and Practice, Oslo, Norway.

Stern N (2007) The Economics of Climate Change: The Stern Review, Cabinet Office – HM Treasury, London, UK.

Sydney Water (2010) Water Conservation Strategy, 2010–15. Sydney Water, Sydney, Australia.

Syed T H, Famiglietti J S, Chambers D P, Willis J K and Hilburn K (2010) Satellite-based global-ocean mass balance estimates of interannual variability and emerging trends in continental freshwater discharge. www.pnas.org/cgi/doi/10.1073/pnas.1003292107.

Thomson W (1889) Electrical Units of Measurement. In: Thomson W. Popular Lectures and Addresses. Macmillan and Co., London I, pp. 73–136.

United Nations (2015) Transforming our world: the 2030 Agenda for Sustainable Development. A/RES/70/1, 21 October 2015. Resolution adopted by the General Assembly on 25 September 2015, 70/1. United Nations, New York

UN DESA (2014) World Urbanization Prospects: The 2014 Revision, CD-ROM Edition. ST/ESA/SER.A/366). United Nations, Department of Economic and Social Affairs, Population Division, New York.

UN DESA (2015) World Population Prospects: The 2015 Revision, Key Findings and Advance Tables. Working Paper No. ESA/P/WP.241. United Nations, Department of Economic and Social Affairs, Population Division, New York.

UN FCCC (2015) Conference of the Parties, Twenty-first session, Paris, 30 November to 11 December 2015. FCCC/CP/2015/L9.

Veolia Water (2011) Finding the Blue Path for A Sustainable Economy. A White paper by Veolia Water. Veolia Water, Chicago, USA.

Walker G (2011) Models of domestic demand in the UK water sector – science or discourse? Workshop on Water Pricing and Roles of Public and Private Sectors in Efficient Urban Water Management, Granada, Spain, 9–11th May 2011.

Water UK (2015) Industry Facts and Figures 2015, Water UK, London, UK.

Willett K M, et al. (2007) Attribution of observed surface humidity changes to human influence. Nature 449, 710-712 (11th October 2007).

WHO/UNICEF (2002) Global Water Supply and Sanitation Assessment 2000. WHO/UNICEF JMP, Geneva, Switzerland.

WHO/UNICEF JMP (2015) Joint Monitoring Programme country files for estimates on the use of water sources and sanitation facilities. http://www.wssinfo.org/documents/?tx_displaycontroller[type]=country_files.

3

The Technologies and Techniques Driving Smart Water

Introduction

In Chapter 1 (1.4.1), it was noted that smart water systems utilise fast (ideally real-time) collection of information, the transmission of this data from its collection point to an interpretation point, the effective collation and interpretation of this data and the presentation of the data in a manner that enables the user to make appropriate decision from this and to be able to respond to the information in a beneficial manner. Ideally, this creates a negative feedback loop. The more data is collected and analysed across different conditions, the greater the system's utility, accuracy and its potential predictive ability.

A resilient water distribution network requires network data in real-time to diagnose failure of supply or network failures such as bursts and to provide integrated control in order to mitigate the impact of any of these in as near to real-time as possible. This is about the proactive as opposed to the reactive maintenance of assets and preventing any actual impact occurring, or minimising and mitigating any impact if it does occur. For example, i2O Water (i2owater.com) specialises in monitoring network valves. Realtime data about how valves are performing enables an operator to diagnose valve and pump condition, assess the potential impact of any assets needing repair or replacement, prioritise service and repair and to predict impending valve or pump failure (Burrows, 2015).

3.1 From Innovation to Application – The Necessity of Integration

In hierarchal and interlinked approaches such as smart water, the technology and its application is only as good as its weakest link and therefore the optimal integration of the separate elements involved plays an important part.

The small size of the water sector in relation to Cleantech and other utilities has meant that in many cases, the enabling technology was originally developed for another application. So, from probes to microbes, smart water systems can involve technology spill-overs from other sectors.

In this case, smart systems have been adapted either by those working in the water sector looking for suitable devices that have been developed in other sectors or by those

Smart Water Technologies and Techniques: Data Capture and Analysis for Sustainable Water Management,
First Edition. David A. Lloyd Owen.
© 2018 John Wiley & Sons Ltd. Published 2018 by John Wiley & Sons Ltd.

who have established smart devices in another sector who are looking for means of increasing their product's sales reach into new segments. This is not necessarily a concern since the cross-fertilisation of ideas and innovations can be a crucible of beneficial progress. It is however necessary to ensure that these approaches have the flexibility to deal with the constraints and challenges which characterise many aspects of water management. Afterthoughts that cannot be fully adapted to address new applications can involve an unacceptable degree of compromise.

The water sector has been characterised by the relative lack of integration of its systems, especially where individual innovations are added to the existing network. For a physical network, this may have a limited impact in terms of perceived service delivery. However, poor integration results in assets across a network operating at less than their optimal efficiency. This may result in more assets being deployed than are in fact necessary, with higher operations and maintenance costs as a result. The effect of this may be more pronounced as assets age and require refurbishment or rehabilitation.

Data processing for utilities is vulnerable to any such shortcomings because of the large number of customers and regulatory obligations involved. Two recent cases in the UK power sector highlight this. Npower has had to pay £26 million to customers and charities after implementing a SAP-based billing and complaints handling system in 2011–13 caused problems for 600,000 customers in 2013–14 (Ofgem, 2015). Similar problems affected 300,000 ScottishPower customers after the implementation of a fully integrated customer IT system in 2012–14 and £18 million in payouts were ordered (Ofgem, 2016).

Poor integration may also make a network vulnerable to perturbations such as population or climate change. For example, a poorly integrated drainage system where the interconnection of rainwater and foul water networks are not properly understood is likely to be vulnerable to causing illegal discharges under more extreme rainfall. For data collection, transmission, processing and its interpretation and display, effective integration is often essential for the system to operate effectively. While this is usually addressed by project developers, the International Telecommunications Union (ITU, 2014) has noted that a lack of common communications protocols remains a concern for the development of smart water. This is explored in more detail in Chapter 9.

The telecoms and electricity sectors have traditionally been more open to innovation than water and sewerage. In the case of the former, this has arisen from competition (especially in long distance, data and mobile services) that has arisen first in North America and Western Europe since the 1980s. In the latter case, this has been due to the need to manage dramatic changes in consumer demand brought about for example, by major televised events and through consumer appreciation about the size of their electricity bills and the need to address them.

As noted in Chapter 1, proof of concept for smart approaches has already been widely demonstrated in the energy sector. This is now increasingly the case for water and wastewater. Most of the technology that will be used until at least 2035 is already being demonstrated in some form. What is less clear is how these technologies will be applied, how they will work with other technologies and which will enjoy the greatest acceptance and adoption.

Integration and ensuring that a system is most efficiently connected and deployed is concerned with how the various components interconnect; the connectivity process and how data is collected and processed (Koenig, 2015). Interconnection is concerned with the connection between an asset and/or a gateway and the cloud (or similar data storage system) for processing and analysis of large amounts of information for operations and maintenance work. Connectivity covers the localised or peer-to-peer (P2P) connections between assets, sensors and systems operators. These include plug-and-play secure wireless connectivity, asset-to-asset communications, asset-to-human communications and asset-to-gateway communications. Data collection and processing (also called at the edge processing) involves collecting relevant data and analysing it in real-time for the effective command and control of the system. Effective interconnection depends on an open and secure architecture that supports developer communities for further application development. This calls for a common language for data collection and analysis which is as flexible as possible through using core services as basic building blocks for larger systems.

Sensors alone do not make a network smart, what matters is their effective deployment and use (Driessen, 2014) as part of a smart network. A truly smart network is one that deploys the minimum number of sensors needed at strategic points across a network for acquiring the necessary real-time data and combining this with available internal and external data for genuinely proactive network monitoring and management. In other words, using the minimum assets and getting the maximum benefits from them.

An example would be the Darwin Sampler developed by Bentley Systems (bentley.com). The Darwin Sampler enables a utility to place the optimal number of monitors within a network for effectively measuring for example, pressure, flow and chlorine levels when modelling the network's performance (Wu, 2015). These applications can be implemented via the Darwin Optimisation Framework which is designed to enable various optimisation approaches to work together at single and multiple locations (Wu et al., 2012), with a Darwin Calibrator for network model calibration.

The Darwin Sampler system is designed to detect old and hard to find leaks as well as finding new leaks. Using loggers is appreciably more efficient than sweeping through a network using acoustic sounding for leaks (Zheng, 2015). This approach also ensures monitoring is continual rather than periodic, so ensuring that new leaks can be swiftly dealt with. This is discussed in more detail in Chapter 5.7.2.

3.2   Digital Manufacturing – The Right Size at the Right Price

In recent decades, digital manufacturing and allied innovations have lowered the cost of manufacturing the ICT components used in smart water systems, along with miniaturising these components and significantly lowering their energy consumption.

This is making it feasible to place smart and smart-enabling elements within systems where in the past they would have been too costly or bulky to be viable in relation to the sector's underlying economics. The improvement in energy consumption per unit of information transmitted also enables hardware to be installed in remote locations without the human and hardware cost of having to regularly change their batteries.

Hagel et al. (2013) outline some examples of the change in cost and power of various ICT technologies and applications in recent years. In computing, the cost of the equivalent processing power of a million transistors has fallen from $222 in 1992 to $0.06 by 2013 and the cost of static random-access memory fell from $44,037 per megabyte in 1980 to $52 per megabyte by 2010 (Brant and O'Hallaron, 2010). For data storage, the cost per gigabyte has decreased from $569 in 1992 to $0.03 in 2012, with the cost of Internet bandwidth in the USA declining from $1,245 per megabyte per second in 1999 to $23 in 2012 and $0.94 by 2014 (Norton, 2014). Finally, between 1959 and 2009, the price of ICT hardware, adjusted for inflation and quality, fell by 16% per annum. It has continued to fall since then, although at some point in the future further progress will slow or cease (Nambiar and Poess, 2011; FRBSL, 2014).

3.3 Smart Objects and the Internet of Things

It is estimated that there were 12.5 billion smart objects in 2010, rising to 25 billion in 2015 and a forecast 50–75 million by 2020. This equals an adoption rate five times greater than the rate of evolution of electricity or telephony (Menon, 2015).

The forecast outcome is that in effect, everything will be connected to everything, and the IoT will become an essential element in new infrastructure. It is clear that some innovations can change consumer behaviour and how it is catered for both rapidly and on a large scale. Menon (2015) highlights the transformative effect of the iPhone, which was launched in January 2007. The Apple iPhone analogy is quite useful as it demonstrates the potential for developing appliances and applications that consumers did not realise that they subsequently felt they had to have.

According to Cisco (Menon, 2015) the Internet of Things will generate revenues and savings of $19,000 billion between 2013 and 2022; $14,400 billion for the private sector and $4,600 billion for the public sector. This is forecast to come from five areas: [1] revenues for innovative devices ($3,000 billion); [2] savings from improving the customer experience ($3,700 billion); [3] the more effective utilisation of assets ($2,500 billion); [4] improved employee productivity ($2,500 billion); and [5] more efficient supply chain logistics ($2,700 billion). IDC (IDC, 2014) believes that the Internet of Things generates revenues of $2,290 billion in 2014, including $100 billion in utilities, which will, grow to $4,590 billion in 2018 with utility revenues of $201 billion, with the actual market devices, connectivity and services growing from $656 billion in 2015 to $1,700 billion by 2020 (IDC, 2015). These forecasts have their own definitions about what the Internet of Things covers and are subject to the commercialisation and adoption of many products and services that are currently at the development stage. As noted in chapter 1.9.2 and 1.16, the Gartner Hype Cycle and Parker Curve suggest that widespread adoption can take 10–20 years from introduction, so these numbers and timescales are perhaps a pointer towards what may happen.

It is the author's belief that many water utilities and regulators are becoming more open to innovation than has been the case in recent decades. This is partly due to an appreciation about how matters have changed in the telecoms and electricity sectors as well as due to the need to address the diverse challenges that water and sewerage utilities face. This will be explored in more detail in Chapter 8.

3.4   The Hierarchy of Smart Hardware and Software

A smart water network was defined by SWAN Forum as having four data layers and a layer of devices that lie outside the actual smart network as outlined in Chapter 1.4.1.5. The operating elements within each of these layers can be summarised as follows (Peleg, 2015; Diaz, 2015), in this case using the urban water distribution network as an example:

1) Automatic decisions and operations.
2) Data management and display.
3) Collection and communication.
4) Sensing and control.
5) Relevant aspects that exist outside the smart network, as the physical layer.

3.4.1   Automatic Decisions and Operations

Areas for data presentation and decision making optimisation include: active leak control prioritisation (what leaks need to be repaired in which order, along with pressure management to minimise ongoing water loss), benchmarking (comparing zones within a utility and comparing various utilities), network anomaly (events taking place beyond accepted performance limits), burst awareness (real-time alerting), burst pinpointing (effective location for remediation) and works optimisation (balancing works between their impact and the amount of work needed to deal with them).

A particularly important area here is the efficient use of data from as wide a range of sources as necessary to inform the operator how events are evolving and how they are anticipated to develop. Integrating weather, river flow, soil saturation and geomorphological data can maximise the time available to respond to a potential flood hazard for example. Another area is in the blending of operational data with historical data to assist in predicting when an asset may need attention. This applies both for alerting when an asset is in danger of failing as well as for making longer-term decisions about prioritising maintenance, refurbishment and replacement programmes. Predictive approaches are typically based on building up information drawing on prior experiences and incidents and refining the monitoring data through feedback loops. As the extreme rainfalls and subsequent floods in the UK during the 2015–16 winter demonstrates, smart decision making also needs to be able to accommodate for unprecedented conditions.

3.4.2   Data Management and Display

There are three principal elements for displaying data: basic reporting and visualisation (information customised so that it is presented in the way that the customer wishes to see and at the level desired, from the entire utility to the individual operational zone); a data repository for current and historic data; and a telemetry system for handling incoming data.

Data displays are being enhanced by being able to put complex information on a screen in an immediately comprehensible form, so that operators know what is happening and fast through geo-located information, presented in layers and with enhanced redlining of critical data. This is illustrated in the Northumbrian Water case study (3.5.1) below.

Major developments since 2005–10 (Heath, 2015) have seen these displays migrate from control rooms to mobile displays, allowing field workers access to graphic displays of data services such as GIS (geographic imaging systems) and CRM (customer relations management). This assists in providing near real-time responses to customer reports of supply problems by fully informing the relevant people who are in the area affected about what action is needed. This reflects the growing impact of mobile data communications and high quality compact visualisation systems. A current concern is that the energy requirements of field units lower their operational effectiveness. Low energy touch screens that are for example being developed by Brodie Technologies are aiming to obviate the need for daily smart mobile device recharging (Peiman et al., 2014).

Recent technological developments have also included the improving resolution of displays. More naturalistic graphics and typefaces mean that they are easier to look at, allowing an operator to absorb information and act on it in a more effective and consistent manner. Another emerging area is the adoption of touch screens, giving the operator greater choice when seeking to interact with information or to go through layers of data. The manner in which information is presented needs to be flexible to that it can be modified to suit each system's circumstances and to be able to prioritise the data presented and the manner in which it is presented. A balance is also needed between the ease of modification by operators and the overall data management requirements along with the utility's priorities.

Data needs to be relevant to the intended recipient. This is particularly important in smart domestic meter displays, where its ease of use and applicability to the customer's needs is central to their acceptance.

3.4.3 Collection and Communication

The best form of data transmission depends upon local circumstances such as what assets are available which may be piggy-backed on, as well as any challenges posed, in ensuring all the data needed can be transmitted securely and in a timely manner.

Adopting communications systems is also driven by how much data is being transmitted and how regularly it needs to be collected. If a utility feels that it needs data on an infrequent basis (collecting water meter readings every few months) the drive-by approach may be the most practical, even though this is not a fully smart approach. As with the migration from non-smart to smart, there is a danger here of creating stranded assets or indeed systems if (or when) a higher frequency of data collection becomes desirable.

Using an extant hard-wire network makes sense when it has already been installed for another service such as electricity or telecoms and that the appropriate utility will allow access to the network at an attractive price. The Internet of Things may also provide additional opportunities for communications via telephone and electricity networks. Otherwise, data communications are made though mobile data services. Four principal services are currently being used: GSM short message service (SMS, a GSM protocol for the exchange of short text messages); GSM general packet radio service (GPRS, a GSM protocol for exchanging data at a medium bandwidth and speed); Radio (dedicated radio transmission services); and Mesh radio (WMN, wireless mesh networks, where a number of transmitters are used to convey data). Using WMN, when one element breaks down or its signal is obstructed signals can be transmitted via the other operational nodes.

3.4.4 Sensing and Control

For an urban network data needs include flow loggers, the smart customer meter, pressure loggers, noise loggers and noise correlators. Other data such as water quality, chloride levels and turbidity can also be gathered using probes, with smart sensors delivering this data in real-time (Heath, 2015). Smart meters are a subset of smart networks and are only as smart as the network they operate within. To use smart meters merely as a faster source of data collection is wasting what they have to offer.

The functions needed depend on circumstances and what needs to be monitored. One of the challenges facing smart water in the future will be to effectively manage the data that is potentially available. Adding parameters for the sake of completeness may in fact simply add an unnecessary degree of complexity to a system. For example, monitoring for potential contaminants will be looking at quite distinct parameters when considering coastal waters, inland waters and the water distribution network.

The list of sensing requirements below is not intended to be comprehensive. Outside the urban network, it reflects areas of current and emerging concern for meeting European Union environmental and public health standards and initiatives in the USA for dealing with irrigation water scarcity and ameliorating potential water use conflicts. In some cases, the effective frequency of testing may (for now at least) not require on-line data collection, although such an approach works well for example with waste incineration facilities sharing emission data with their regulators on a real-time or near real-time basis in the EU for at least the past 25 years.

Coastal waters: Bathing water quality (faecal contamination), CSO overflows and coastal levels and flows (flood monitoring).

Inland waters: Water flow (volume and height, for flood warning and management, resource planning and inland water quality), temperature (water quality, via the ability to retain dissolved oxygen), colour and turbidity (underlying contamination issues and individual pollution incidents, along with flooding and exceptional surface run-off), contaminants (faecal contamination, pesticides, and heavy metals, for example, from agricultural, urban, industrial and mining discharges and also for water leaving a service reservoir), pH and dissolved oxygen (eutrophication).

Water treatment: Water flow (in and out of the facility), contaminants (before, during and after treatment), pH (impact on plumbosolvency), turbidity, residual disinfectant and operational information including energy use, dosing and asset performance and condition.

Water distribution: Water flow (day and night for leakage assessment), water pressure (by district management zone), pump operation and performance, contaminants (for example, bromate, cyanide, mercury, various pesticides, Clostridium and total organic carbon) and information needed to monitor the condition of the network.

End user: Contaminants (testing at customer tap zone, as outlined by the World Health Organization's Guidelines for drinking-water quality, 4th Edition, 2011), water flow (via a smart water meter), sewage flow (via a smart wastewater meter), and monitoring systems at the device level (via the Internet of Things).

Sewerage: Foul water and storm water flow through the foul and storm networks respectively and through combined sewerage networks, sewerage network condition and capacity along with pump operation and performance.

Sewage treatment: Pump operation, effluent flow, energy use, asset operation and condition, effluent parameters (before, during and after the treatment process), input of treatment chemicals, waste products (at each treatment stage, by volume and appropriate quality criteria).

Wastewater recovery: Energy generated (along with energy use at the facility and energy exported), nutrient concentrations (nutrient recovery), contaminants, colour and turbidity (water recovery). This includes the flow, input of chemicals, throughput of wastes and recovered materials, pump operation, and asset performance and condition.

Effluent discharge: Contaminants (nutrients, BOD_5, potentially also certain pharmaceuticals, along with antibiotics and oestrogen, temperature (for process water), and flow.

Irrigation: Water (and nutrient) input, soil moisture levels, sap flow and nutrient concentrations.

Information gathered within the network may be blended sensing data collected outside the network (weather, social media and energy prices for example) as well as applicable data from other smart networks, along with (where appropriate) historic (legacy) data.

3.4.5 Relevant Aspects that Exist Outside the Smart Network, as the Physical Layer

A mentioned before, device that is not connected to a smart network is by definition, not a smart device. All sensors, probes and meters are devices for gathering information. It is the transmission, interpretation, display and application of the information that these appliances gather which is smart. This includes the sensors in a district metering area, the bulk meter and the customer meter.

A similar hierarchy of data gathering, transmission and application applies for areas such as smart irrigation, river monitoring, flood monitoring and management, and sewerage and sewage treatment management.

3.4.6 Smart Water Grids as Integrated Data Hierarchies

The smart water grid offers another hierarchy of data gathering, with information flowing through a number of smart water systems. At the simplest level, this concerns the various data flows for a city or utility from water resources, treatment and distribution, to its use by domestic and other customers and its flow, treatment and discharge through the sewage system (Mutchek and Williams, 2014). Further layers of information can be added including direct (clean water) and indirect (post-use water) irrigation by various end users, water quality and characteristics at the points of abstraction and discharge, and natural water flows through the utility area. These data flows can be integrated for various outcomes such as the potential vulnerability to flooding both from surface water and sewer overflows or to relate various forms of water use to available water resources.

On a larger scale, Singapore's Public Utility Board (PUB) is developing a smart water grid for integrating how the city-state's water resources (rainwater, desalination, water reuse and water imports) are most effectively integrated in terms of energy and operating costs as well as ensuring the security of its supplies (PUB, 2010). This is in turn being integrated with distribution network pressure management and metering to optimise water consumption and inland water monitoring to ensure suitable water quality for the reservoir network. Korea aims to develop a '3S' (Security, Safety Solution) national smart water grid platform by 2020 (Choi and Kim, 2011) with a network of micro-grids (serving individual towns and cities) feeding into macro-grids (regional or river basin level) and ultimately providing a national overview of water data. By linking all applicable data flows, including dam water levels, river flows, agricultural usage and water and wastewater treatment, the platform is intended to mimic the water cycle cross the country. This will be reconsidered in Chapter 8.

For industrial customers, a closed-loop approach towards water management is becoming increasingly popular, as it allows a facility to minimise both its water consumption and its effluent discharges, thereby maximising the economic value created from each unit of water consumed. The Korean initiative is to some extent, seeking to replicate this, albeit on a fundamentally different scale.

3.5   Case Studies: Towards Implementation

The five case studies below survey the various ways smart water applications have been developed for utilities. These include demand management, generating and using big data, automation and efficiency savings. These themes will be considered in greater detail over subsequent chapters.

3.5.1   Case Study 3.1: Northumbrian Water's Regional Control Centre

3.5.1.1   Northumbrian Water's Aims and Outcomes

Northumbrian Water (nwl.co.uk) has been adopting the foundations of smart approaches since the 1980s. In 1980, basic telemetry was introduced, along with retrospective monitors for looking at past data later that decade. From 1990 district metering areas (DMA) were adopted, along with integrated network management. From 2000–10, a hydraulic model was developed, Netbase (for the analysis of social media) introduced along with water production planning and a new regional control centre in 2007.

The regional telemetry system was replaced in 2010 with an integrated system covering the entire utility. This allowed connection and operation with NWL's other corporate systems with an emphasis on improved analysis. The new Regional Control Centre involved the replacement of telemetry system (SCX2 to SCX6.7 eScada) and the adoption of two new software systems (Hawkeye and Aquadapt). There were historic issues from inadequate alarm handling capability and weakness with the legacy SEEK (data management) system.

A SCX Server acts as the telemetry and incident database, working with the utility's external supporting database (Oracle/SQL Server). It takes in data from ViewX Clients (Auto-Dialler Interface) and SMS Interface (send/receive).

In 2002 NWL had an annual production plan (PP) for the operating expenditure. The PP process went to a quarterly cycle in 2005, but this had a minimal impact. In 2007, the PP moved to a weekly review cycle which impacted operations on a daily basis. This encouraged NWL to adopt a consistent daily PP review in 2008 for network optimisation and a process that was not exposed to subjective evaluation.

Aquadapt was installed in 2015. The Aquadapt system is for the real-time control and optimisation of the water distribution system. Under Aquadapt, schedules of treatment works throughput, treated water pumps, control valves, service reservoirs and towers are updated every 30 minutes. This means there is a constant review about the cheapest water resources to use, with a focus on energy costs. It is linked to SCADA (SCX6.7 eScada) and outstations and the degree of automation means that manages in effect become supervisors, albeit with the ability to take direct control at any time this is needed.

The facility is now the central repository for time-series data from telemetry and data loggers, offering network connectivity with schematics and analysis tools along with visualisation and spatial analysis of network and asset data through flexible reporting and customised views. It includes a configuration tool that allows data cleansing and maintenance in-house. Outputs include district metering area leakage calculations and reporting and hydraulic model management.

The enhanced control and supply/demand management has led to a 6% reduction in water produced, a 14% improvement in energy efficiency and a 20% reduction in energy consumption. This has resulted in a £1.7 million sustainable annual reduction on operating expenditure.

In terms of performance, the Production Plan has been accurate to within 1–2%, while Ofwat service indicator metrics (SIM) are showing a consistent Upper Quartile performance amongst the ten Water and Sewage Companies, with the lowest leakage in the sector for Essex and Suffolk Water (NWG South) and the 4th lowest for Northumbrian Water (NWG North), along with the best performance for interruptions to supply for the past two years.

The company's experience to date suggests that successful implementation is based on making the best use of what smart data has already been generated and bringing in new smart data where it adds value to what is already being used.

Smart Analytics is enabling NWG to identify deviations and changes in network performance that impact service at an earlier stage, which means there are less network events due to human error, as well as network optimisation and the improved delivery of service outcomes for customers and regulators. The next phase involves Smart Field Operations, the moving of decision-making capabilities into the field. NWL is seeking to ensure that staff capacity is built to compliment system capacity. These data were adapted from Austin and Baker (2014) and Baker (2015).

3.5.1.2   Smart Systems for Northumbrian Water – Schneider's SCADA

Schneider upgraded Northumbrian Water's ClearSCADA (SCX6.7 eScada). NWG's requirement was to link its telemetry/SCADA control room with rotas, contact management, telephony, work logs, and knowledge capture. The utility's Action Management System (AMS) is designed for lowering the costs of alarms to improve traceability and communications while reducing re-typing and errors. This is carried out by defined business processes and uses rules to determine actions and advise the operator about all alarms received. The AMS determines fault type, looks up skills data and checks staff availability using a series of rules and the database. The AMS was developed through an 'agile' time-bounded contract based on specified key features and its implementation and specification developed through weekly workshops bringing developers and users together. (Adapted from Beadle, 2015.)

3.5.1.3   Smart Systems for Northumbrian Water – Aquadapt's Water Management System

The evolution of the Aquadapt water management system demonstrates the time and complexity involved when developing innovative offerings. Derceto (derceto. com) started to develop Aquadapt water management software in 1997. It was originally a one-off consulting engineers' project carried out with the client and involved 800 hours of project development. The second project in 1999 involved 2,000 hours of project development and had a $200,000 budget for a product with SCADA for database and display. Again, this was a one-off project, this time with more limited client input.

In 2004, Aquadapt became a product in itself, requiring a further 14,000 hours of development work. As it was no longer a one-off project, a structured query language (SQL) database with replication was used, with the user interface now being separate from the SCADA and open platform communications (OPE) standards for communications. A proper integrated development environment (IDE) was used, with an automated development and testing environment and for the first time, comprehensive record keeping.

The Aquadapt Echo evolved from these products. A one model fits all approach was adopted, using the principle of agile development. On the 15th version of the model, a satisfactory system was developed. From this point, revisions are locked in once they are stable and care is taken to track developments and to be able to roll them back when needed. A FogBugz issue tracking system has been incorporated. Everything is tested, with a development loop for the prototype, looking at its design and functional requirements with the client. The product was then further developed through testing and final testing with the client until it was satisfied. A Continuous Integration process involves a daily in-house upgrade to latest version of all client databases and build of help files and installers.

After 55,000 hours of product development over 13 years, two principal products have been rolled out; the Historian Database and the Live Database. Marketing Aquadapt depends on bridging perceptions; between what customers seek (low cost) and what is achievable (not a cheap product to develop). Individual projects typically take 4,000–7,000 hours to deliver for each client, reflecting the feature-heavy nature of the system.

The Aquadapt Live Database has an operator panel with a degree of customisation offered allowing the new user to change the graphics to merge with their current systems. Network pressure and leakage data is managed with relation to operating decisions such as energy needs (and energy prices) and water demand expectations against resources while generating operating schedules and monitoring alarms. Data is provided chiefly at the district metering zone level and then aggregated into larger areas. There is a SCADA interface (with the water utility SCADA) and functions for energy pricing, which interface with the Historian Database's dashboard via the Internet/ Intranet. The Aquadapt Historian Database interacts with Live Database for data replication. It has a dashboard (energy pricing interface) and an application manager (which also interacts with Aquadapt Live Database), data archive, database query function and a strategic operations simulator.

In 2015, there were 21 installed systems, all of which are kept within two revisions of the latest version as part of a continuing product development process. The idea being that no change can break anything for any client.

Derceto had to write off the 55,000 hours of initial project development work, as subsequent business has yet to make up for this. At the same time, the French water utility and technology company Suez (Lyonnaise des Eaux) has found it difficult to develop a suitable smart water strategy for metering in France. In 2010 the company chose to upgrade its metering systems internally after tension between the utility and its supplier. As a consequence, in 2014, Suez acquired Derceto in order to use Aquadapt for its smart network management and to further develop the product. (Adapted from Bunn, 2015 and Perinau, 2015.)

3.5.2 Case Study 3.2: Big Data at Dŵr Cymru Welsh Water

Dŵr Cymru Welsh Water (DCWW, dwrcymru.com) serves approximately three million people for water and sewage services in Wales and parts of England. It was privatised in 1989 and spun out of Hyder Plc in 2001 as Glas Cymru Cyf a privately owned, not for profit company.

DCWW is adopting a 25-year smart water strategy which is planned to run from 2010–35. DCWW's smart data strategy started in AMP 5 (2010–15) with full implementation during AMP 6 (2015–20) and it is anticipated that AMPs 7–9 (2020–35) will be about the effective integration of the systems and the information they generate. Their aim is to provide an invisible customer service (ensuring that their customers do not notice the background work being carried out to deliver their services), to minimise field work to what is actually needed in the right place and at the right time and to maximise the productivity of any field work. The utility has gone as far as they can on improving conventional operating efficiency, especially since the Glas Cymru reorganisation in 2001. As a private sector company, the management needs to make a business case for each smart application; there is no room for experimental development at this point in the company's evolution.

DCWW anticipates big data playing a significant role in 2020–25, along with the connected home and the Internet of Things emerging by that time. Currently DCWW generates 331,500,000 data points per annum, including 180 million for water flow, 53 million for wastewater flow, 36 million for telemetry outstations, 47 million for water pressure and 7 million for business customers. This data is managed by two people during the day and one at night at its monitoring headquarters, along with three regional centres, each being staffed by one person. Their experience to date demonstrates that when you have a small team of managers, what matters is the important data, not the noise. (Adapted from Bishop, 2015.)

3.5.3 Case Study 3.3: Non-Revenue Water Reduction at Aguas de Cascais

Aguas de Cascais (aguasdecascais.pt) serves 208,000 people through the largest water concession in Portugal, generating €38 million pa in revenues. Since the 30-year concession started in 2000, non-revenue water (NRW) has been one of the main areas of concern. NRW was reduced from 39% in 2001 to 25% in 2005 by conventional NRW approaches. NRW was kept stable until 2011 (25.6%) against a national median of 30.7%. In 2010, NRW was costing the utility €2.98 million pa, equivalent to 8% of revenues.

The second NRW reduction phase involved increasing the frequency of leak repairs from 500 per annum in 2009 and 2010 to 2,200 in 2012, through implementing an active leak control programme in June 2011. The leak detection teams were incentivised by guaranteed work throughout the year and a management policy of translating leakage data is into revenue data, so water leaks are in fact seen as fiscal leaks, involving Euros per hour, as well as physical leaks in terms of cubic meters of water per hour.

The third phase involved implementing smart approaches. All data analysis was automated from 2012 and the district management area size was reduced and pressure management systems brought in. Leakage data is overlaid on Google Earth and sent to leakage management teams. From this, a leakage database is assembled, including the normal volume of water consumed and the maximum noted, along with the duration of the leakage from its detection to completion. Each event is given a unique identity and linked to the relevant before and after event data.

Large night-time consumers have full two-way AMR metering to assist in distinguishing between their usage and the background usage levels. Current loss is calibrated against the potential minimum physical loss and the minimum practically achievable physical loss.

Leakage management has also moved towards minimising the time taken for dealing with each leak and improving the quality of the work to maximise the pipe's effective operational life. The complete repair time (locate the leak, stop the water flow, repair and resume water supplies) has been decreased from 4 hours and 54 minutes in 2011 to 3 hours and 56 minutes in 2014. Improved understanding of network performance by 2014 has allowed the utility to move the active leak control programme towards optimising network pressure management, adopt smart leakage detection approaches and develop a predictive pipeline and asset management programme based on reducing background leakage. This involves the selection of pipes each month for rehabilitation or replacement (Table 3.1).

The overall impact of the programme has been to move the utility from being an average performer in to one of the better performers, especially in terms of the leakage per kilometre of pipeline, as shown in Table 3.2.

Other areas being developed include more rapid and effective customer response and lowering energy usage through energy efficiency measures and automating energy consumption data analysis. Water meter data is also indirectly used for measuring the water going into the sewerage network. The approach here is to consider all sources of data

Table 3.1 Impact of the NRW control programme, 2010–2014.

Year	NRW
2010	26.6%
2011	25.6%
2012	17.3%
2013	15.1%
2014	14.3%

Adapted from Perdiago (2015).

Table 3.2 Aguas de Cascais, NWR performance in 2012 and 2014 (national data is for 2010).

NRW	Percentage	m^3/km pa	m^3/connection pa
Best	8.7%	1.381	33
Median	30.7%	2.653	75
Worst	41.0%	8.813	180
A de C 2012	17.3%	2.537	82
A de C 2014	14.3%	1.870	60

Adapted from Perdiago (2015).

that can be sensed as sources of information which in turn can be monitored to seek further ways of improving efficiency in the future.

In 2014, the cost of NRW was €1.248 million against €2.977 million in 2010. After the costs associated with the programme, this is generating a net saving of €1.5 million per year (Perdiago, 2015).

3.5.4 Case Study 3.4: Smart Meter Services for Aguas de Portugal

The Portuguese regulator has set efficiency targets for utilities allying these to financial incentives. Nationally, 40% of municipal water is lost or unbilled in 2011, equating to €200 million pa, which fell to 30% in 2015 (GWI, 2017). The aim is to reduce NRW to 25% by 2025 under the PNUEA national efficient water use programme government plan as set out in the ERSAR – RASARP 2012 (Portuguese Regulatory Report).

EPAL (Aguas de Portugal, adp.pt) is a government-owned utility that provides a bulk water supply to 34 municipalities to manage water supplies in Lisbon. EPAL serves a total of 550,000 consumers in Lisbon, or 347,000 individual customers via 84,500 connections.

In seeking to lower its NRW, EPAL set out five targets. [1] Identify and quantify the actual water flow through the networks through DMA network segmentation and continuous telemetry monitoring. [2] In order to quantify how much water is lost, generate basic data about the clients and the network, including its length, number of connections and pressure. [3] Decide what the priority areas are through developing selection criteria and performance indicators and analysing relevant data. [4] Locate where water is being lost for the optimisation of active leak control. [5] Identify where repairs are needed and carry them out rapidly and well.

A four phased plan was rolled out to enable EPAL to understand its network performance:

Phase 1: DMA planning and set-up. Create metering points and telemetry systems, design and boundary validation and DMA implementation. This involves network segmentation and continuous telemetry monitoring. Boundaries and monitoring points, with locked in DMAs with closed boundary valves. 156 DMAs were created, monitoring approximately 1,250 km of networks, and 98% of clients. There are a total of 1,600 monitoring points including 350 network metering and telemetry systems and over 1,000 client telemetry systems.

Phase 2: Monitoring systems. Continuous monitoring of pressure and flow, with passive data collection systems linked to active alarms for anomalies.

Phase 3: Data analysis. Through developing simple and effective data analysis systems based upon leak detection targets and results validation. This generates huge volume of data that needs to be integrated through appropriate software to link it to the relevant performance indicators.

Phase 4: Reporting of information. This extends from the definition of proposed DMAs to analysing how each DMA is performing both on its own and in relation to other DMAs and auditing their performance and generating audit reports.

3.5.4.1 EPAL's DMA Analysis Project Methodology

EPAL's aims were to create a system of DMAs which could be subject to continuous telemetry monitoring allied with simple and effective data analysis systems for leak detection target definition, the optimisation of active leak control activities and results validation. The focus has been to address what is essential and to control the utility's costs.

Desk-based work included a detailed DMA performance review of net nightline water flow and authorised use analysis to estimate recoverable losses and for nightline target setting. Fieldwork was based on a 'find and fix' approach including DMA boundary valve validation, leak detection and correlation and ground microphones, with the data recorded on a GIS, temporary DMA alterations, leak repairs and validation of the subsequent results.

The water optimisation for network efficiency (WONE) data integration platform was mainly developed in-house to meet EPAL's practical requirements, interfaced with other management systems and uses data management and performance ranking. WONE includes DMA telemetry and strategic customer telemetry data, the SCADA system, GIS, G/InterAqua (a water and wastewater asset monitoring and management system developed by AQUASIS, a subsidiary of EPAL), and the AQUAmatrix client management system (also developed by EPAL) for meter readings, billing and customer support.

The principal dashboard gives the most important data and system management information, including highlighting the poorest performing DMAs and the highest avoidable losses, along with providing a daily summary of the overall picture and the total avoidable losses at the network level. At the DMA level data is interpreted through graphic displays highlighting any areas of concern in particular regarding leak detection, quantification and repair verification. This data is presented via pressure and flow profile graphs for each DMA to identify where the critical point pressure can be improved.

Understanding how assets work in reality only occurs when you are in command of information about significant yet easily overlooked events. For example, a major aquarium is cleaned every Tuesday morning, which was creating a ten-meter pressure drop in one DMA which can now be factored into that DMA's pressure management schedule.

3.5.4.2 Implementing Innovation

EPAL found that the most effective way of implementing and managing a programme such as this was by provoking cultural change across the utility, both in terms of personnel management and data management. This depends on staff training and development as well as understanding how the network operates in reality. This in turn required a dedicated team of staff working on NRW reduction with full management support and adequate resources to complete the task. It is also evident that further resources are needed to maintain the progress and that the cost savings made from the programme justify these efforts.

3.5.4.3 Results to Date

Between 1993 and 2001, NRW varied between 19% and 23% with distribution losses of 37–46 million m^3 pa (Donnelly, 2013) (Table 3.3). The work carried out since 2010 builds upon previous programmes using more traditional technologies.

The improvement in the network's ILI was greater than for volumetric and percentage terms. This demonstrates that ILI is a particularly effective measure as network efficiency improves. The fact that the network has a 'B' rating suggests there is scope for further gains despite what might appear to be an impressive 7.9% NRW figure in 2013. NRW was 8.5% in 2015, the second lowest in Portugal (GWI, 2017).

The programme has resulted in consistent improvement across all indicators, with 100 million m^3 of water saved between 2005 and 2014. Lessening EPAL's water needs has other benefits, including saving €0.8 million in chemicals used, €5.5 million savings in energy used with overall savings of €7 million over this period. It has also allowed the utility to rationalise and defer investments, resulting in improved business resilience and easing the need to raise tariffs.

Table 3.3 EPAL, non-revenue water, 2002–13.

	Million m^3 pa	% NRW	ILI
2002	32.04	25	Not assessed
2005	26.93	23.5	11.1 (D)
2010	13.97	11.8	5.7 (C)
2013	8.17	7.9	3.1 (B)

Adapted from Donnelly (2014).

Looking at some DMA examples shows the degree of losses than can be identified. DMA 'A' has 2.6 km of network mains and 800 clients. Its assessed background losses were calculated at 0.5 m^3 per hour with an assumed client usage (allowance) of 1.2 m^3 per hour. In fact, a flow of 18 m^3 per hour was detected; 1.7 m^3 per hour being accounted for and 16.3 m^3 per hour unaccounted for. DMA 'B' has an 8.5 km network and 2,150 clients. Here the 'inevitable' losses were 1.2 m^3 per hour, with a customer allowance of 3.9 m^3 per hour. The actual flow was found to be 8.0 m^3 per hour, 2.7 m^3 per hour being unaccounted for.

In the two most extreme cases, water flow was reduced from 110 m^3 per hour to 50 m^3 per hour, equivalent to 1.4 million m^3 per year. The savings from these two DMAs in effect paid for EPAL's entire DMA project in three years.

As well as improved performance, it is clear that this approach has lowered operating costs, especially through improved network management and control from a better understanding about how the network performs. It is also clear that network performance can be unpredictable, with a series of significant leaks in a DMA occurring after a long period when no problems were noted. EPAL has made cumulative savings water of around 120 million m^3 between 2005 and 2014 (Donnelly, 2014 and Donnelly, 2016).

3.5.4.4 The Waterbeep Service at EPAL

Smart meter deployment formed part of EPAL's WONE project. In order to appreciate the performance of the DMAs it was important to take into consideration the consumer profile of EPAL's larger customers. Smart metering therefore started with EPAL's 900 largest customers and is being rolled out in subsequent phases.

The Waterbeep service allows customers to monitor their own consumption based upon pre-defined field alerts that the customers can select and adjust to their own requirements. It has been designed for four separate market segments and in turn allows utilities to manage and transfer their customer profiles. This reflects the realisation that customer behaviour affects the way a utility operates and therefore the need for companies to continually adapt to new demands and requirements.

Four levels of smart metering service are offered:

1) Basic, primarily for home users, and free of charge.
   This service monitors average water consumption and compares these with the average values in the city. A complimentary smartphone app has been developed for communicating meter readings. This can be used by all customers and can be enhanced by the customer taking further meter readings themselves.
2) Plus, costing €1 per month.
   This level allows the customer to measure their consumption over a set period via telemetry data, as well as in the last 30 days and 7 days. The customer receives an alert via text or e-mail when their consumption is outside set parameters, which can be customised. This is aimed at domestic and small business customers.
3) Pro, costing €12 per month.
   This level provides consumption data for every 15 minute for a set period of time such as the previous day. It is aimed at large businesses. (12€/month)
4) Premium, costing €20 per month.
   This enables allows the integration of water use data within a company's own systems.

Consumption data is provided to customers through a web application that aggregates and combines information from different CRM systems and telemetry. Alerts are calculated based on reference consumption values and are sent by text message or e-mail whenever consumption is different from the usual profile.

By providing information to customers about their water usage patterns, it is helping them to be more alert to deviations in their water consumption. This makes saving water a routine activity since it provides customers with information about their average consumption and alerts the utility about consumption peaks associated with domestic and network leakage. The data and the way it is presented assists customers to identify atypical consumption and to avoid large bills that usually result in complaints and less satisfaction.

One consideration is the need to ensure that the bills make sense to customers while providing them with all the relevant data that they might need. Waterbeep was launched with a promotional campaign, supported with media and outdoor advertising, leaflets, web and social media publications aiming to reach all their domestic, commercial and industrial customers (Branco, 2014).

3.5.5 Case Study 3.5: The Vitens Innovation Playground

Waterbedrift Vitens (vitens.nl) is a state-owned utility providing water to 5.6 million people in the Netherlands. The Vitens Innovation Playground (VIP) in Leeuwarden covers 750 km^2 in six DMAs for monitoring water flow, pressure, temperature, conductivity, and quality for 100,000 households and 2,270 km of network mains. A total of 106 sensors are involved; 23 flow monitors, 23 pressure monitors, 15 for conductivity and 45 Eventlab water quality sensors (see Optiqua below). It is one of four test beds being used by the EU supported SmartWater4Europe programme. Other demonstration sites are managed by Acciona Agua (SWING, Burgos), Thames Water (TWIST, Reading), and the Université of Lille (SUNRISE, Lille). This is a €12 million, four-year programme looking at lowering energy use by 10–15%, leakage detection and localisation, real-time water quality monitoring and improving customer interaction.

3.5.5.1 Performance and Practicalities

Customer calls can become a source of information rather than simply as a means of dealing with queries, concerns and complaints. For this, a utility can monitor calls for their meaning (are they telling us something we ought to know) and by mapping where the calls, what this can show.

Proactive customer management means translating realtime data capture into anticipating where problems may occur. Vitens holds customer e-mail, social media and mobile phone data and aggregates this into specific areas so that when an event occurs, it can inform customers by e-mail, SMS and social media before their customers are aware that there is a problem. This has resulted in a reduction of call centre contacts by 90% and improved customer satisfaction, as they are now more engaged with what has been going on.

3.5.5.2 The Beginnings of Big Data

Vitens anticipates dealing with 120 million values per day across their entire network coming from their sensors alone. Other data inputs that need to be taken into account and integrated with these include, social media (especially Twitter), incoming calls (complaints, alerts and queries), traffic data, special dates that may affect demand, trending topics (using Google), weather, and sensor data from other sources.

Examples of companies working with Vitens and SmartWater4Europe reflect the diversity of opportunities that are emerging through the assistance of a supportive utility.

3.5.5.3 Incertameter

Vernon Morris and Co (founded in 2007) developed Incertameter (incertameter.co.uk) for instant leak detection and data transmission units based on the more precise and timely measurement of water depth and pressure. These were initially trailed with Yorkshire Water. They can also be used to generate alerts of imminent storm sewerage overflows.

3.5.5.4 Quasset

Quasset BV (quasset.nl) is a Dutch company founded in 2011 specialising in asset integrity management and condition assessment for utilities and petrochemical facilities through the integration of sensing, analysis and simulation.

3.5.5.5 Optiqua

Optiqua (optiqua.com) is based in Singapore. Its EventLab uses a sensor to measure changes in water refractivity that are brought about by contaminants. These changes are compared with baseline data to flag any abnormalities and to alert the operations system about them. Field trials with PUB Singapore and Vitens indicate that the EventLab 2.0 system is capable of detecting toxins at concentrations significantly below those required by the World Health Organization guidelines.

3.5.5.6 Arson Engineering

Arson Engineering (arson.es) is part of Spain's Grupo Arson, which dates back to 1975. It is developing smart water data transmission and integration systems for a variety of urban applications, including houses, offices and industrial sites under the Aquacity system. Trials are being carried out in Bilbao, Cuidad de Burgos (with Acciona Agua) and La Pobla de Farnals (with Aguas de Valencia) as well as with Vitens. Other subsidiaries include ArsonMetering (smart meter management) and AquaArson (irrigation management through integrated control of irrigation systems).

3.5.5.7 Scan Messtechnik GmbH

s::can Messtechnik (s-can.at) is an established manufacturer of online optical water quality monitors with revenues of €10 million pa. In 2012 it launches the i::scan compact LED based sensor for the continual real-time monitoring of water quality and potential contamination, looking at turbidity, total organic carbon, colour, pH, chlorine, conductivity and other parameters throughout the water network. These are offered in a nano::station unit.

3.5.5.8 Homeria

Homeria Open Solutions (homeria.com) was spun off from the University of Extremadura in Spain in 2008 specialising in data management and modelling for smart cities. It has developed management user interfaces for areas such as monitoring water consumption and allowing the manager to simulate outcomes under a variety of different scenarios.

3.5.5.9 StereoGraph

StereoGraph (stereograph.fr) is a French company founded in 2006 for developing immersive three-dimensional representations of buildings and facilities allowing developers to better understand how they will function before they are built.

3.5.5.10 Mycometer

Mycometer A/S (mycometer.com) is based in Denmark, offering rapid microbiological testing systems. It has been operating since 2004 with revenues of €2 million pa. The company is currently developing an online microbiological water quality testing system designed to be used across the water distribution system to enable managers to have near real-time alerts for potential contamination incidents.

These data are based on Driessen (2014), Van den Broeke (2015) and Thijssen (2015).

Conclusions

Smart water has been driven by the development of smart approaches towards electricity management, especially smart grids and demand management. It has also benefitted from the reduction in size, cost and energy needs of hardware used for data collection, transmission and assimilation. While originally an adjunct of smart energy management, smart water has emerged as an umbrella expression to cover a number of approaches that have emerged in their own right. Systems integration and the development of the Internet of Things will result in various self-sufficient approaches being interlinked. This offers the potential to optimise the incremental advantages each single approach offers into a coherent overall approach. The greater complexity and the data generated by such an integrated approach will bring its own challenges, especially in terms of security and managing the data to best reflect actual needs rather than delivering externally assumed information outcomes.

Applications and their implications will be explored in greater detail in the following chapters; Chapter 4 will consider smart domestic water approaches and Chapter 5 will explore the potential to optimise water and wastewater management at the utility level.

References

Austin A and Baker M (2014) Smart networks: Our journey towards leading network performance. Presentation to The 'Smart' Water Network, CIWEM, London, 4th December 2014.

Baker M (2015) Vendor or in house solutions: Trade off or opportunity? Presentation at the SWAN Forum 2015, Smart Water: The time is now! London, 29–30th April 2015.

Beadle S (2015) A vendor perspective. Presentation to the SWAN Forum 2015 Smart Water: The time is now! London, 29–30th April 2015.

Bishop M (2015) Presentation to the SWAN Forum 2015 Smart Water: The time is now! London, 29–30th April 2015.

Branco L (2014) Waterbeep – smart efficiency. Presentation to the SMi, Smart Water Systems Conference, London, 28–29th April 2014.

Brant R E and O'Hallaron D (2010) Computer Systems, A Programmer's Perspective. Pearson, 2nd Edition (p. 584).

Bunn S (2015) A vendor perspective. Presentation at SWAN Forum 2015 Smart Water: The time is now! London, 29–30th April 2015.

Burrows A (2015) Intelligent control: optimal and proactive networks. Presentation at the SWAN Forum 2015 Smart Water: The time is now! London, 29–30th April 2015.

Choi H and Kim J A (2011) Alternative Water Resources and Future Perspectives of Korea. Presentation to the 2011 IWA-ASPIRE Smart water Workshop, 4th October 2011, Tokyo, Japan.

Diaz, E M (2015) Remarks made at the SWAN Forum 2015 Smart Water: The time is now! London, 29–30th April 2015.

Donnelly A (2013) WONE water optimisation for network efficiency. Presentation to the SWAN Conference, London, 23–24th May 2013.

Donnelly A (2014) Water optimisation for network efficiency: Applying effective tools for reducing non-revenue water within a major utility. Presentation to the SMi, Smart Water Systems Conference, London, 28–29th April 2014.

Donnelly A (2016) Modern methods of managing non-revenue water. Presentation at Potable Water Networks: Smart Networks, CIWEM, 25th February 2016, London, UK.

Driessen E (2014) Smart Water Grid: Smart meters do not make water grids smart… Presentation to SMi, Smart Water Systems Conference, London, 28–29th April 2014.

GWI (2017) Portuguese municipal operators face €235 million NRW bill. GWI weekly news, 23rd March 2017.

Hagel J, Brown J S, Samoylova T, Lui M, Arkenberg C and Trabulsi A (2013) From exponential technologies to exponential innovation. Report 2 of the 2013 Shift Index series. Deloitte University Press/Deloitte Development LLC.

Heath J (2015) Smart Networks. Presentation to the SWAN Forum 2015 Smart Water: The time is now! London, 29–30th April 2015.

IDC (2014) Worldwide Internet of Things Spending Guide by Vertical Market 2014–2018 Forecast. IDC, Framingham, MA, USA.

IDC (2015) Worldwide Internet of Things Forecast, 2015–2020. IDC, Framingham, MA, USA.

ITU (2014) Partnering for solutions: ICTs in Smart Water Management. ITU, Geneva, Switzerland.

Koenig M J (2015) Intelligent connectivity: The fabric of a smart water system. Presentation to SWAN Forum 2015 Smart Water: The time is now! London, 29–30th April 2015.

McIntosh A (2014) Partnering for solutions: ICTs in Smart Water Management. International Telecommunications Union, Geneva, Switzerland.

Menon A (2015) Merging smart cities with smart water. Presentation at SWAN Forum 2015 Smart Water: The time is now! London, 29–30th April 2015.

Mutchek M and Williams E (2014) Moving towards sustainable and resilience smart water grids. Challenges 5: 123–137.

Nambiar R and Poess M (2011) Transaction Performance vs. Moore's Law: A Trend Analysis. Berlin, Germany: Springer.

Norton W B (2014) What are the historical transit pricing trends? DrPeering International
http://drpeering.net/FAQ/What-are-the-historical-transit-pricing-trends.php.

Ofgem (2015) Notice of intention to impose a financial penalty pursuant to section 30A(3)
of the Gas Act and 27A(3) of the Electricity Act 1989. Ofgem, 18th December 2015,
London, UK.

Ofgem (2016) Notice of intention to impose a financial penalty pursuant to section 30A(3)
of the Gas Act and 27A(3) of the Electricity Act 1989. Ofgem, 26th April 2016,
London, UK.

Peiman H P, Wright C D and Bhaskaran H (2014) An optoelectronic framework enabled by
low-dimensional phase-change films. Nature 511: 206–211.

Peleg, A. (2015) Water Networks 2.0 – The power of big (wet) data. Presentation to the
SWAN Forum 2015 Smart Water: The time is now! London, 29–30th April 2015.

Perdiago P (2015) A smart NRW reduction strategy. Presentation to the SMi Smart Water
Systems Conference, London, April 29–30th 2015.

Perinau T (2015) Comments at the SWAN Forum 2015 Smart Water: The time is now!
London, 29–30th April 2015.

PUB (2010) Water for all: Meeting our water needs for the next 50 years. Public Utilities
Board, Singapore.

Thijssen R (2015) Water quality…from a customer perspective. Presentation to SWAN
Forum 2015 Smart Water: The time is now! London, 29–30th April 2015.

Van den Broeke J (2015) The case for real-time water quality monitoring. Presentation at
SWAN Forum 2015 Smart Water: The time is now! London, 29–30th April 2015.

Wu Z Y (2015) Integrating data-driven analysis with water network models. Presentation
to the SWAN Forum 2015 Smart Water: The time is now! London, 29–30th April 2015.

Wu Z Y, Wang Q, Butala S, Mi T and Song Y (2012) Darwin Optimization User Manual.
Bentley Systems, Watertown, CT 069795, USA.

Zheng Y W (2015) Integrating Data-Driven Analysis with Water Network Models SWAN
Forum 2015 Smart Water: The time is now! London, 29–30th April 2015.

4

Domestic Water and Demand Management

Introduction

Domestic and municipal users are the smallest consumers of water globally after agriculture and industry. They also are often the most high profile consumers because of the higher price charged for their water and sewerage services, customer expectations about public health and service delivery standards and the challenges involved in supplying a substantial number of people living and working in a small area in relation to locally available water resources.

Smart domestic water metering and demand management tools are a distinct segment of smart water. There is a significant overlap with smart water networks in assisting the utility to appreciate water flows at the district metering level and for monitoring water balances. This chapter considers domestic smart metering from both a customer and a utility perspective.

4.1 Metering and Smart Water Metering

Before customer behaviour can be influenced and modified, customers need access to information about their water usage, the cost (and other) implications of this usage and how it can be beneficially changed.

Water metering is the foundation of demand management. Demand management is driven by the information it generates. The more information each meter generates, the greater the potential to influence consumer behaviour. This is because when water billing is not related to customer usage, there is no incentive for the customer to consider their consumption.

4.1.1 Adoption of Metering

Water metering is neither universal nor universally accepted. While there is a broad trend towards adopting water metering, there is no guarantee that it will be broadly adopted. This is despite its evident merits. Table 4.1 presents a global overview of metering penetration by country. The metering rates relate to the percentage of domestic customers formally served by a water utility. Data is either the latest available or at or after the point where it has already reached universal (100%) coverage. In many cases, actual universal coverage was achieved quite some time earlier than this.

Smart Water Technologies and Techniques: Data Capture and Analysis for Sustainable Water Management,
First Edition. David A. Lloyd Owen.
© 2018 John Wiley & Sons Ltd. Published 2018 by John Wiley & Sons Ltd.

At the national level, there are some notable contrasts in the metering adoption data highlighted in Table 4.1. Urban Australia is typically water scarce, while that is of appreciably less concern in New Zealand. Universal urban metering is seen in all the South East Asian countries, despite differing degrees of economic development. Metering is the norm in Europe with the exception of the UK, Ireland and Norway. As will be shown in Table 4.2, it is becoming the norm in England and Wales, while it is not being adopted at present in Northern Ireland and Scotland. Ireland aims to have universal metering by 2017. By May 2016, 820,000 out of 1.05 million customers in Ireland had meters installed (Kelly, 2016), but their actual use has been put on hold due to a change of Government resulting in proposed water charges being suspended. In both Northern Ireland and the Irish Republic, there was a tradition that water is supplied free. Scotland has appreciably greater per capita water resources than England and Wales. While water services are fully privatised in England and Wales, they are state held in Scotland and Northern Ireland.

4.1.2 The Adoption of Metering in England and Wales

The expansion of metering in England and Wales from 0.1% in 1989 to 43% by 2015 along with the tentative adoption of smart metering within this timescale along with the high quality of data available covering this process means that it deserves attention. Before privatisation in 1989, the only significant domestic water metering in England and Wales was at a trial starting in 1988 covering 53,000 properties in the Isle of Wight by Southern Water (Smith and Rogers, 1990). Since 1990, customers in England and Wales were allowed to opt for a water meter ('optants') and they have to be installed on all new builds. In recent years, Ofwat has become more amenable towards metering and this has been reflected by Southern Water's programme during AMP5 (2010–15, see Case Study 4.4 at the end of this chapter).

As shown by Table 2.25 in Chapter 2, the installation rate has increased consistently over the first two decades (1990-2010) from 7.150 million in 1990–00 to 10.275 million in 2000–10 and this accelerated in 2010–15 with 7.395 million meters installed during the half decade. During these 25 years, a total of 24.820 million meters were installed. AMP6 (2015–20) will see this progress maintained with major programmes at Thames and Yorkshire resulting in more than half of domestic customers having a water meter.

What happens when metering starts to become the norm? In England and Wales, unmetered properties are billed according to their rateable value in 1989. Therefore a single person living in a large house would pay more for water than a large family living in a small property irrespective of their water consumption. Before 2005, households either had water meters because there was a clear benefit in having a meter installed (optants) or because they lived in a new build.

Historically, consumption has increased. For example, unmeasured consumption at South West Water rose from 108 litres per person/day in 1977 to 130 by 1990 (Hooper, 2015), then to 154 in 1997–98 and 171 by 2003–04 (Ofwat, 2005), before levelling off at 173 by 2014–15 (Water UK, 2015). Metered consumption shows a different trajectory, first rising from 122 in 1997–98 to 141 in 2003–04 (Ofwat, 2005), and then falling to 119 in 2014–15 (Water UK, 2015).

Table 4.1 Metering penetration in single-family houses with a water supply (%).

	Year	Penetration		Year	Penetration
Western Europe					
Austria [2]	1998	100%	Belgium [2]	1997	90%
Denmark [2]	1996	64%	Finland [2]	1998	100%
France [2]	1995	100%	Germany [2]	1997	100%
Greece [2]	1998	100%	Ireland [6]	2015	0%
Italy [2]	1998	>90%	Netherlands [2]	1997	93%
Norway [2]	1998	<20%	Portugal [2]	1998	100%
Spain [5]	2010	92%	Sweden [2]	1998	100%
Switzerland [2]	1998	100%	UK (E&W) [4]	2015	44%
UK (Scotland) [4]	2015	1%	UK (N Ireland) [4]	2015	0%
C&E Europe					
Bulgaria [1]	2014	98%	Czech Rep [1]	2013	100%
Hungary [1]	2007	100%	Poland [1]	2015	100%
Romania [1]	2010	92%	Slovakia [1]	2007	100%
Rest of Europe					
Croatia [1]	2004	82%	Russia [1]	2014	69%
Ukraine [3]	2004	35%			
MENA					
Egypt [1]	2010	85%	Jordan [1]	1998	100%
Kuwait [1]	2010	91%	Tunisia [1]	2010	100%
Turkey [2]	1998	>90%			
S-S Africa					
Cote d'Ivoire [1]	2014	98%	Kenya [1]	2014	90%
Mozambique [1]	2014	78%	Nigeria [1]	2014	13%
Senegal [1]	2014	96%	South Africa [1]	2014	91%
Tanzania [1]	2014	98%	Uganda [1]	2013	100%
North America					
Canada [1]	2005	61%	USA [2]	1997	>90%
South America					
Argentina [1]	2014	21%	Brazil [1]	2014	84%
Chile [1]	2006	98%	Colombia [1]	2010	93%
Mexico [1]	2012	91%	Peru [1]	2014	67%
Venezuela [1]	2006	38%			

(Continued)

Table 4.1 (Continued)

	Year	Penetration		Year	Penetration
SE Asia					
China [1]	2013	100%	Cambodia [1]	2013	100%
Japan [7]	2015	100%	Korea [1]	2013	100%
Malaysia [1]	2007	100%	Philippines [1]	2009	100%
Singapore [1]	2013	100%	Vietnam [1]	2009	100%
S and C Asia					
Bangladesh [1]	2015	84%	India [1]	2005	58%
Kazakhstan [1]	2014	73%	Pakistan [1]	2010	3%
Tajikistan [1]	2005	42%			
Oceania					
Australia [1]	2007	100%	New Zealand [1]	2015	58%

Adapted from: [1] IB-Net (www.ib-net.org); [2] OECD (1999); [3] OECD (2007); [4] Water UK (2015); [5] Iagua (2010); [6] Kelly (2016) and [7] JWRC (2016).

The rise at South West Water may have been exceptional, but the divergence between metered and unmetered consumption is seen across England and Wales. While average unmetered water consumption for England and Wales has been in a range of 139–161 litres per person/day between 1991–92 and 2013–13 (and 148–160 between 1995–96 and 2013–14).

Metered consumption fell from 153 in 1991–92 to 118 in 2013–14 with a range of 115–133 between 1995–96 and 2013–14 (Hooper, 2015). Table 2.26 compared the difference at five yearly intervals. In 1999–00 and 2004–05, unmetered customers used 9% more water per capita. In 2009–10, the difference was 18% and 19% in 2014–15 as metering started to significantly influence customer behaviour (Ofwat, 2000; Ofwat, 2005; Ofwat, 2010; Water UK, 2015).

Once adoption moves on from the most obvious beneficiaries (the author's water bill was halved when he moved house and became an optant in 1996), it would be expected that the difference between metered and unmetered consumption to decrease as meter penetration increases. In fact, the reduction in water consumption was 9% in 2000 and 2005 but 18–19% in 2010 and 2015. While unmetered use does not appear to have changed much during this period, there is a fairly consistent decline in metered consumption. Given that more of the higher usage customers are being metered, this would suggest that consumer behaviour is being modified by metering.

Likewise, it would be expected that the areas with the lowest metering would show the biggest consumption reductions (Table 4.2). Again, there is no significant relationship between metering levels and water reductions with metering. Except for one crucial element, where meter penetration is above 60% (two WaSCs and two WOCs), the reduction is 14–15%. The ability to afford water intensive white goods (unmetered households, along with their impact on electricity bills) let alone use them (metered households) is likely to be a driver for low water usage at Yorkshire (metered customers

Table 4.2 Metering in the Utilities in England and Wales, 2014–15.

	Metering rate	90%+ metering	Metered impact	Serious stress	Roll-out	Smart meters
Affinity Water	75%	2025	15%	Yes	Compulsory	AMR
Anglian Water	70%	2020	31%	Yes	Optional [1]	Trial
Bournemouth Water	60%	2025	10%	No [6]	Optional [1]	No
Bristol Water	66%	2030	15%	No [6]	Optional [1]	No
Cambridge Water [2]	70% [3]	2035	–	No [6]	Optional	AMR
Dee Valley	58% [5]	No	23%	No	Optional	No
Dŵr Cymru Welsh	40%	No	28%	No	Optional	No
Essex and Suffolk [4]	59% [3]	2035	–	Yes	Compulsory	No
Northumbrian	44%	No	6%	No	Optional	AMR
Portsmouth	25%	2030	27%	No [6]	Optional	AMR
Severn Trent	39% [3]	No	–	No	Optional	No
South East	65%	2020	17%	Yes	Compulsory	No
South Staffs	36% [3]	No	–	No [6]	Optional [1]	AMR
South West	78%	2040	31%	No	Optional [1]	No
Southern	85%	2020	14%	Yes	Compulsory	AMR
Sutton and East Surrey	46%	2040	28%	Yes	Optional [1]	AMR
Thames	34%	2030	18%	Yes	Compulsory	AMI
United Utilities	38%	No	25%	No	Optional [1]	AMR
Wessex	60%	2035	10%	No [6]	Optional [1]	No
Yorkshire	48%	No	31%	No	Optional [1]	AMR

[1] Compulsory for change of occupier of a property under varying circumstances.
[2] Cambridge water and South Staffordshire meter reduction data has been combined.
[3] 2013–14 data.
[4] Essex and Suffolk now report within Northumbrian Water.
[5] 2012–13 data.
[6] Classified as Stressed in the original 2012 draft.
Sources: Percili and Jenkins (2015); Priestly (2015); Water UK (2014); Water UK (2015); Individual company 2015–2040; Water Resources Management Plans (2014).

using 31% less water) and UU compared with high usage in more prosperous regions such as those covered by South East and Affinity Water.

Metering priorities have been influenced by external events. In 2004–05, Thames Water has 5% metering and anticipated this rising to 40% by 2029–30 and 80% by 2034–35 (Godley et al., 2008). As a result of a drought in 2006, this was brought forward effectively full coverage (in excess of 90%) by 2029–30 (Thames Water, 2014). While there is no specific support for smart metering by Ofwat, allowing utilities in areas of extreme water stress to compel customers to use water meters since 2008 does mark a significant change in approach by the regulator.

Table 4.2 summarises the current state of play in England and Wales. Unless otherwise stated, the metering rate is for 2015. A point above 90% (typically 93–97%) is regarded as the cost effective limit for meter installation. The impact of metering is shown by the difference in water consumption between metered and unmetered properties in 2014–15. In England and Wales, after some 15 years of installing traditional meters, demand was 9–21% lower after optant meters were installed and 10–15% for compulsory meters (NAO, 2007). The Environment Agency classified six companies as facing serious water stress in their 2012 assessment, which were downgraded to not serious in the final 2013 assessment (EA/NRW, 2013). As of 2013, seven of the 19 utilities were classified as facing serious water stress, with five adopting a compulsory metering policy, and ten installing or trialling AMR technology and one (Thames Water) currently installing AMR meters designed to be upgraded to AMI. This is set to evolve. For example, in June 2016, Anglian Water announced that it is to trial 7,500 smart meters with Sensus and Arqiva with domestic and non-domestic customers in and around Newmarket, Suffolk over a four-year period. The trial seeks to optimise customer satisfaction, eliminate bursts or leakage and to reduce water consumption to 80 l/c/day. Meter readings will initially be made once an hour, and are set to increase to once every 15 minutes later in the trial (WWi, 2016).

It is evident from Table 4.3 that there is a wide variation in average (metered and unmetered) domestic water consumption and a further wide variation between consumption reduction targets. There is no evident relation between consumption reduction targets and metering targets. The Environment Agency regards 130 l/c/day as a realistic consumption target within this period. This does appear to be conservative, given that metered customers in seven of 17 water utilities in England and Wales used less than 120 l/c/day in 20-14-15 (UK Water, 2015).

Table 4.3 Domestic consumption – 25-year water resource management plans.

Litres per capita/day	2015–16	2039–40	Reduction
Anglian	130	114	12.3%
Severn Trent	129	117	9.3%
Cambridge Water	140	125	10.7%
South Staffs	137	128	6.6%
Bournemouth	152	132	13.2%
Affinity	169	139	17.8%
Bristol	159	140	11.9%
Portsmouth	157	149	5.1%
South East	163	149	8.6%

Source: Based on Engineer, 2015.

Table 4.4 City tariff structures as surveyed by GWI.

Tariff	2010	2011	2012	2013	2014	2015	2016
Increasing	141	163	165	177	185	199	207
Increasing then decreasing	0	3	3	3	3	3	3
Linear	122	144	145	146	145	149	155
Decreasing	6	11	11	11	12	11	11
Fixed	4	6	5	6	6	4	4
Free	3	4	4	4	4	4	4
Total	276	331	333	347	355	370	384

Source: Adapted from GWM 2011, 2010 and GWI, 2016.

4.1.3 Tariff Structures

Global Water Intelligence has carried out a global tariff survey for towns and cities since 2007. It is reasonable to assume that to be surveyed by GWI, the utilities will serve substantial towns or cities and to have relatively sophisticated operations. This is reflected in the fact that 2–3% of those surveyed either charge only fixed fees or no fees at all. In the case of utilities with partial metering (such as in England and Wales) the metered tariffs are used.

Table 4.4 has been developed from the municipal tariff surveys for cities globally round the world by Global Water Intelligence since tariff structures were included on a systematic basis in 2010 (GWM 2011, 2010 and GWI, 2016).

Overall, 93–95% of the utilities either had linear or increasing (rising block) tariffs. There was a small towards increasing tariffs (from 51% to 54%) at the expense of linear tariffs (44% to 40%).

4.2 Types of Water Meter

There are three general types of water meter. The traditional mechanical (or 'dumb') meter, which measures the water flow continually and meter reading are taken via a physical inspection. Automated Meter Reading (AMR) which do not require a manual inspection and Advanced Metering Infrastructure (AMI), where AMRs are in effect integrated into a data collection and processing network. The latter are smart meters.

Data is logged in three ways. The traditional accumulation meters log up how much water has been used since the previous reading (typically every three, six or 12 months). A pulse meter records the time taken for a certain volume of water (100 litres for example) has been consumed and provides a readout of these time intervals. An interval meter records how much water has been consumed over a given period of time (an hour or a day, for example). The interval meter operates continually, while the pulse meter is only activated when a given volume of water has passed through. Both pulse and interval meters can give more data through smaller set volumes and time intervals respectively. The finer the data, the greater the scope for its interpretation.

Meters themselves have evolved. Unlike a traditional paddle meter, electromagnetic flowmeters and ultrasonic meters do not come into contact with the water. The lack of moving parts means they have a significantly longer operating life and lower maintenance costs but they do require a power source. Since smart meters by definition require a power source that is not a specific concern.

4.2.1 Types of AMR Meter Reading

Data still has to be collected from an AMR meter, either by a hand-held unit taken to the property or using a remote unit which is driven past the property. Hand-held reading, is either touch based, where the meter reader connects with the meter via a probe or when the reader is sufficiently close to the unit while walking by a series of meters. Unlike accumulation meters, a considerable amount of data can be transmitted from a single AMR visit, rather than a single reading.

In drive-by systems, an operator drives to an area and downloads meter data from the meters within its transmission range. Each meter has its own identifier code, so there is no need to visit properties in any sequence.

4.2.2 Smart Metering – From AMR to AMI

AMR reading is appreciably faster than for traditional meters, especially when drive-by is used. It also allows for more data to be collected. But this data is only collected periodically, no matter how detailed the actual data is, and there is no interaction between the meter and the customer.

Smart meter systems typically use remote data transmission so that the utility receives information at a frequency that would be difficult if not impossible to achieve by visiting the meters, even when using drive-by reading due to the staffing costs involved. Wireless systems use a variety of protocols to transmit data to utilities and where desired to users.

While the original AMRs were concerned with automated meter reading data transmission, AMRs have subsequently been developed to allow a wide variety of data to be transmitted, including activity patterns and leakage alerts, meter and battery condition and the flexibility to enable billing systems to suit specific customer needs. It is the connection to the communications network that makes the difference here and these are correctly regarded as smart meters and work with an advanced metering infrastructure (AMI), hence the meters being referred to as AMIs.

More advanced AMI approaches use a home area network (HAN) which allows meter data to be sent to the user as well as the utility and full two-way communications, where in addition, the utility can communicate with the meter and the customer via the HAN. Two-way communication using for example digital cellular networks allow meters to be remotely updated. As well as minimising the need for visits to update the unit's software, this is particularly important in maintaining system security and integrity.

Communications can be via established mobile data networks and electricity networks on a data sharing basis (piggy-back), minimising the need to duplicate equipment. In more remote locations, the HAN may use satellite communications. In some urban applications, dedicated data transmission networks are used.

One important difference between smart electricity and water meters is that the former has a readily available power supply. As a result, it is often assumed that a smart water meter will transmit one reading or a set of readings once a day or less often, in order to maximise battery life. There are circumstances where more frequent transmissions will be used.

4.2.3 Smart Water Meters and Demand Management

Smart metering is the most recognisable and well-known aspect of demand management. While smart metering has a somewhat limited role in smart water overall, it can be seen as smart water's public expression, as it is the aspect that members of the public are by some degree most likely to encounter. Smart metering is designed to modify customer behaviour through informing them about the explicit link between their water consumption and their water bills to minimise both, along with the implicit link between water and energy bills.

Smart metering in practice covers the periodic measurement of water flow and the transmission of this information to users and utilities. The customer interface should provide data about water consumption and its cost on at least a daily basis. Further levels of information can include forecasting bills, comparing usage with peer groups, anomalous water consumption alerts, the ability to remotely shut the water supply, and data at the device level.

Rising block tariffs in theory ought to encourage demand management as water tariffs are banded on how much water is consumed each month or year. This is also designed to improve affordability as water consumed for essential use (hygiene, cooking, etc) has a lower unit cost than water for non-essential use (gardens, car washing and swimming pools, etc), so that the more non-essential water is used, the higher its marginal price. Human nature also matters; where rising block tariffs are used (Millock and Nauges, 2009) there is evidence that customers 'game' their water usage to the upper edge of each tariff level. This creates an incentive to maximise water usage within each tariff block.

4.2.4 The Cost of Smart Metering

While a traditional meter is a stand-alone unit, a smart meter requires a communications infrastructure in order to operate. This is the reason why smart water metering systems usually cost appreciably more than traditional meter deployments.

According to Thames Water, in the UK, a traditional meter has an average 60 years net present value (NPV) cost of £580 per connection, against £630 for AMR meters and £750 for an AMI fixed network. As NPV benefits rose with the metering technology used, from £80 for traditional meters to £400 for AMI, the more advanced systems are the most cost-effective, albeit with NPV costs £350 higher than the benefits. Many of the benefits come from the role smart metering plays in demand management at the river basin management plan level (Slater, 2014).

The data is based on the cost per average property. The AMI example includes 16-year contracts for meter installation and management to 2030 for meters installed in 2015. There are also further cost benefits when it is possible to have a synchronised roll-out with AMI for electricity utilities, especially in sharing the communications infrastructure. In addition, there is the improved ability to appreciate cross-cutting savings through common data platforms for the customers (Slater, 2014).

Table 4.5 Cost breakdowns for smart water metering infrastructure in the USA.

$ per household	East Bay [1]	Various [2]	Santa Barbara [3]
Meter	80	102–163	111–155
Transmitter	75–100	6–11	6–32
Box lid and connector	20	–	23
Installation	70	34–89	44–59
Collection network	20	12–23	1–24
Total	260–285	154–286	204–287

Adapted from [1] EBMUD (2012), [2] Beecham Research, cited in Sierra Wireless (2014), and [3] Westin (2015).

A study by Arqiva, Artesia and Sensus in England and Wales was summarised in Table 2.27. It sought to quantify how technology selection might impact water companies, consumers, and the environment (Hall, 2014).

The dramatic difference between AMR and AMI for water and carbon savings is due to the difference in consumption between traditional and AMR metering, which is low, compared with the further savings which are attained via AMI services. These are summarised in Table 4.11 below.

There are a wide variety of estimates for the cost of the various AMI network elements. This in part depends on what is included in each breakdown as well as the number of meters involved in each roll-out. Table 4.5 summarises three cost breakdowns in the USA. It is likely that the roll-outs cited by Beecham Research (Sierra Wireless, 2014) cover more meters than the other two cases.

The smart meter unit itself accounted for 28–66% of the total roll-out cost, the higher proportion being in the larger roll-outs with their lower per household communications infrastructure costs.

Table 4.6 summarises 14 smart mater deployments in Europe, Australia, and North America. It highlights the variety of projects involved, all of which include a degree of retrofitting, whether starting with a traditional or an AMR meter. In the case of Orland, the cost of the AMI infrastructure was the same at $0.61 million for 2,600 meter points, or $233 per meter point, while the difference in per meter costs was driven by the number of meters to be upgraded to AMI. For Halifax, the first figure is for the original tender and the second is what will be actually paid. This highlights the difference between various tenders and quotations and what can in fact be paid.

In Santa Fe, the AMI system replaced an AMR based system installed 10 years before. Small-scale trials and limited deployments such as at Orland will have higher per capita costs than large-scale deployments as seen in Ottawa. AMR to AMI retrofits benefit from the incumbent meter unit typically remaining in place, with the AMR component being replaced by an AMI unit.

The fact that AMR units are being replaced by AMI in seven cases here (three partial replacement and four complete replacement) underlines the transient nature of a technology that offers incremental improvements in data collection and storage without being fully smart. It also raises the issue of stranded assets, where a utility invests in one technology and spends more money replacing this before the end of its economic life.

Table 4.6 Cost of smart meter and infrastructure deployment.

Client	Project	Meters	Per meter
Elk Grove, USA [2]	Dumb to AMI	12,296	$257
Wichita Falls, USA [8]	Dumb to AMI	34,000	$471
Wide Bay, Australia [9]	Dumb to AMI	26,500	A$226
San Francisco, USA [15]	Dumb to AMI	178,000	$337
Orland, USA [4]	Dumb to AMI (replace 10+ yr meters)	2,224	$676
Orland, USA [4]	Dumb to AMI (replace 15+ yr meters)	1,047	$1,051
Jersey, UK [11]	Dumb and none to AMI	36,600	£205
Halifax, Canada [3]	Dumb/AMR to AMI	82,336	C$210
Halifax, Canada [7]	Dumb/AMR to AMI	82,336	C$305
Brunswick Co., USA [12]	Dumb/AMR to AMI	34,041	$257
Orange, USA [5]	Dumb/AMR to AMI	21,240	$282
Cedar Hill, USA [13]	Dumb to AMI	16,000	$563
Santa Fe, USA [14]	AMR to AMI retrofit	36,000	$167
Malta [10]	AMR to AMI, water and electricity	245,000	€163
Port Townsend, USA [1]	AMR to AMI retrofit	4,661	$432

Adapted from: [1] Honeywell (2013); [2] Carey (2015); [3] Halifax Water (2014); [4] Carey (2014); [5] DS&A (2016); [6] M&SEI (2011); [7] M&SEI (2016b); [8] M&SEI (2016a); [9] Waldron (2011); [10] OECD (2012); [11] Snowden (2013); [12] BCU (2015); [13] Hamblen (2016); [14] Miller (2015) and [15] Wang (2105).

4.2.5 Operating Costs for Smart Metering

Smart meter costs lie mainly in data transmission, management and display rather than its collection. The less contact with the meter, the cheaper data collection becomes, as summarised in Table 4.7. While only one item of data can be collected by reading a mechanical meter, a single reading from an AMR can provide an appreciable amount of data. Likewise, a daily AMI reading may in fact contain readings taken at least every 15 minutes.

Table 4.7 Cost of water meter reading, by technology.

€ per reading	Flat	House	Commercial
Dumb – walk up	1.000	3.000	10.000
AMR – touch	0.700	1.200	5.000
AMR – walk by	0.200	0.500	2.000
AMI	0.003	0.003	0.003

Adapted from Sensus, data for Europe, cited in Godley et al. (2008).

Assuming a daily AMI reading, a three-monthly AMR reading and a six-monthly traditional meter reading, on this basis, annual meter reading costs for a house would be €6.00 for a traditional meter, €2.00–4.80 for AMR and €1.10 for AMI.

By comparison, the town of Naperville (Illinois, USA) is covered by a service contract $1.13 per annum for six bi-monthly walk up AMR reads or $0.19 per read (Bookwalter, 2016). In Santa Fe, California, 36,000 AMI meters are covered in a $2 million 10-year service contract, which includes the meters being guaranteed for this period at $5.56 per household per annum (Miller, 2015).

Fathom (gwfathom.com) is an offshoot of Global Water Resources (gwresources. com), an Arizona-based water utility which specialises in supplying water to areas in North America with limited water resources. Fathom demonstrates how AMI generated data both provides cash-flow for AMI specialists and can also create further business through its application.

From a utility perspective, Fathom generates AMI meter revenues from a basic charge of $1.50 per meter per year, appreciably cheaper for the utility than traditional manual or drive by AMR meter reading and covers all standard data requirements. Further revenues arise by encouraging customers and utilities to migrate towards more complex services. As the company has access to all customer data generated by subscribing utilities, this can become a business in its own right. They can offer data comparisons and have the potential for benchmarking between utilities. There are 4 million Fathom meters installed in the USA (Symmonds, 2015).

4.2.6 Smart Meter Deployments to Date

At the end of 2014, there were an estimated 67 million AMR and AMI meters installed in the USA, 42 million AMR and 25 million AMI. There has been a consistent shift towards AMI, with AMI accounting for 8.3 million of the 23.9 million shipments in 2005–09 and 17.0 million of the 28.1 million shipments in 2010–14. (The Scott Report on AMR and AMI Deployments, 2015, quoted in DS&A, 2016). 10 'notable' AMI roll-outs in the USA were noted in 2014 (EBMUD, 2014), covering 2.88 million meters, with 2.46 million having been installed. Five had web interfaces operational and these are being planned for the other five.

A survey of 19 roll-outs and trials in Australia and New Zealand and 15 in the rest of the world that were announced between 2009 and 2013 (Boyle et al., 2013) noted 1.34 million AMR and 0.61 million AMI domestic meters. This can be seen as a starting point in smart metering deployment *(Boyle et al., 2013)*. A total of 20 trials (8–5,000 meters) and 6 roll-outs (more than 10,000 meters) in Australia and New Zealand were active at the end of 2014 against 12 and 5 roll-outs at the end of 2013. These covered 152,000 meters in 2013 rising to 205,000 in 2014 (Beal and Flynn, 2014).

4.2.7 Metering Deployment, Development and Utility Cash-flow

Moves to metering, let alone smart metering face many obstacles. Most utilities operate on a cash-flow basis, so any activity that has the potential to change what were previously carefully managed and anticipated cash-flows from billings will by its nature carry an element of risk. Where no meters are used, a utility has fairly precise expectations about its future cash-flows, since the only variables are the number of properties being billed and what tariff each property will pay.

With metering, the utility does not know to the same degree what its cash-flow will be until it receives the billing data. In a manual read collection system or AMR system, water use data is uploaded by the utility at the end of a billing period and so the data collection remains disconnected from the billing system during each billing cycle. Customers respond to the cost of their consumption at the end each billing cycle.

In AMI networks, water usage data is sent continually to the utility and the customer. This means that customer behaviour can be modified within a billing cycle. As the utility is receiving this data on a continual basis, it can monitor water consumption and cash-flow development within the billing cycle. When consumption and billing data is available in real-time or near to real-time, it becomes more predictable, as a utility will be in control of this data and does not need to make assumptions between formal readings. This is a process of adapting to different data regimens.

As metering systems in a utility are typically rolled out in a series of phases, this means that different parts of a utility's customer base will be at different stages of the adoption process at the same time. This will be the case when moving to no metering to metering of any type, or during the transition from manual or AMR metering to AMI (Symmonds, 2015).

4.3 Smart Metering in Practice

4.3.1 What Data Means for Utilities and their Customers

There is an essential asymmetry in water network data when it comes to customer consumption; utilities can know too much, while customers usually know too little. Utility managers need to differentiate between the mega-data collected and the relevance of the information it provides. The former is concerned with generating information, sometimes with an incomplete appreciation about how it is to be used. In contrast, information is the application of that data in a manner that can benefit its recipient. Mega-data therefore can be of value for both customers and utilities when it is appropriately managed and interpreted. For the customer, this means providing them with information that they can quickly understand and appreciate its value to them.

Appreciating customer behaviour and what motivates it depends on sound evidence. This requires a developing suitable benchmark or baseline for all subsequent comparisons and analysis. That means the utility needs to appreciate where it currently stands with regards to how it understands its customers, what they are going to do which is likely to modify customer behaviour and how their your service will look like when you have completed this (McCombie, 2014).

4.3.2 The Need to Appreciate Customer Behaviour

Every degree of increased customer and utility involvement in consumption and billing data requires both parties to appreciate the practicalities involved, especially with regards to the utility understanding how customers will view these changes. The detailed appreciation of customer behaviour is a recent and swiftly evolving development in water utility management.

Concerns about social norms can be combined with knowledge, awareness and economics in motivating customer behaviour. This be can either carried out through intermittent feedback, advising the customer how much water they consume per day against similar households in their area (peer comparison) on a monthly basis (by letter) or by continuous feedback (via a digital device), where consumption data is updated hourly. In the latter case, information such as how much water is used per hour and comparisons between daily and weekly consumption can also be generated (Javey, 2016).

It is also important to be realistic about how much any given price signal to a customer can achieve on its own (Smith, 2015). Acceptance of metering is more likely to occur when it is part of a broader package of customer-related and water-saving initiatives. Customers are less likely to be impressed by metering schemes where the utility has high water losses. Likewise, customers need to be fully informed about metering as a way of saving both money and water without taking up much customer time. Customers need to be segmented to account for their specific needs, as shown in the Southern and Thames Water case studies (Case Studies 4.4 and 4.6).

Privately owned utilities have appreciably less scope than state-held entities to compel customers to change their behaviour other than to install meters where they are permitted to. This was highlighted in the seasonal tariff trial that was carried out by Wessex Water in 2008–11 (Wessex Water, 2011 and 2012). Beyond this, cooperation and education are needed. For example, 40% of customers surveyed by the Consumer Council for Water (a statutory stakeholder body representing water utility customer interests in England and Wales) in 2008 supported compulsory metering, 25% opposed it and 35% were undecided. 60% regarded metering as the fairest basis for billing, 15% preferred rateable value and 25% were undecided. 60% supported more metering with 20% against and 20% undecided (Lovell, 2016). The Consumer Council for Water also found that more than 50% of customers would reduce their water usage with smart metering systems if usage and price data was made clearly available and price comparisons were provided (Smith, 2015).

4.3.3 Water Metering and Demand Management

Water meters are the enabling tools for demand management. They inform the customer about their water consumption and motivate them to modify their water consumption. Metering has an impact at two levels. A traditional meter informs the customer about their overall water consumption once every three, six or twelve months. The customer may then chose to respond to this by making some broad changes to their water consumption. The same applies for AMR meters, although the customer has more data to base their decisions upon. An AMI meter ensures the customer gets more timely and detailed information about their water consumption, allowing them to see the impact of each individual intervention.

With one exception (the Isle of Wight trial) the examples in Table 4.8 do not strip out the water loss through internal leakage identified by metering. They date from 1988 (Isle of Wight) to 2015 (Southern Water).

These examples point to 10–16% reduction in consumption when traditional meters are used and a further 7–15% reduction when traditional meters are replaced by AMI systems. Where AMI (or in the case of Wessex Water, a device providing AMI level data at the household level) is used with previously unmetered households, 16–17% reductions were noted.

Table 4.8 Impact of metering on domestic water consumption.

Utility	Country	Change	Type	Service
Southern Water [1]	UK	−16.5%	AMI	New
Wessex Water [4]	UK	−17.0%	'Smart'	New
Southern Water – IoW [8]	UK	−10.0%	Dumb	New
England and Wales [2]	UK	−11.0%	Dumb	New
Literature review [3]	Global	−12.5%	Dumb	New
Southwest Water [5]	USA	−16.0%	Dumb	New
East Bay, California [9]	USA	−15.0%	AMI	Replace
Sunnyvale, California [6]	USA	−12.0%	AMI	Replace
Cedar Hill, Texas [12]	USA	10.0%	AMI	Replace
Four trials [7]	Australia	−10–13%	AMI	Replace
Riyadh [11]	Saudi Arabia	−9–10%	AMI	Replace
Dubuque, Iowa [10]	USA	−6.60%	AMI	Replace

Adapted from: [1] Ornagi and Tonin (2015); [2] WSA (1993); [3] UKWIR (2003); [4] Wessex Water (2012); [5] Pint (1999); [6] Javey (2016); [7] Beal and Flynn (2014); [8] Smith and Rogers (1990); [9] EBMUD (2014); [10] IBM (2011); [11] Elster (2010) and [12] Hamblen (2016).

Earlier studies exhibit higher savings. Larger savings in the USA are at least to some extent driven by lower garden irrigation. A survey prepared for the California Urban Water Conservation Council (A&N Technical Services, 2005) noted 13–45% reductions in the USA and Canada in trials carried out between 1946 and 1972, 20–40% in the USA (1958–65), 14–34% in Israel (1970s) and 34% in Malmo, Sweden (1980). This survey does note that that the quality of this data should be seen as being poor (A&N Technical Services, 2005). Poor data quality is at least in part due to a weak understanding of water flows.

A typical AMI deployment here is by Badger Meter (badgermeter.com). In its AMI system, data is collected and transmitted from the household, via an Orion cellular endpoint, to the company's Beacon cloud system where it is transmitted to the utility for processing before being returned to the cloud system and transmitted to the customer via a digital device (EyeOnWater) or by letter (Water Focus Reports). Initial results in the city of Sunnyvale, California found a 12% reduction in demand after the AMI system replaced the traditional meters. Although consumer behaviour modification was the chief aim, a significant proportion of the reduction came from improved customer side leak detection.

In a typical urban water distribution system, 25–30% of distribution losses occur within household boundaries rather than the network. Without metering this usually goes undetected, unless the leak is noticed by the customer. Metering can play a significant role in countering this, through the detection of anomalous water consumption. A further analysis of the Isle of Wight data (Godley et al., 2008) noted that consumption in fact declined by 10–11% due to lower domestic consumption and a further 10–11% from lower internal leakage.

Smart metering will impact internal leakage as anomalous water consumption will be detected earlier, in real-time, rather than through comparing data through periodic reading cycles and due to smaller changes in consumption becoming discernible.

Thames Water has found a wide variation in consumption. In 2013–14, 11% of their customers consumed less than 100 litres per person per day, 61% 100–200, 21% 200–300 and 5% 300, and 2% more than 400 (Nussbaum, 2015).

4.3.4   Multi Utility Metering

Case Study 4.3 notes that demand for electricity is driven down when consumption data is made more immediately available to the consumer. This agrees with the observations made above about metering methods and domestic water consumption. The next step ought to be to inform customers about the impact of water usage on customer energy bills.

An estimated 89% of Britain's CO_2 water-related emissions are generated by domestic water heating or 5% of total greenhouse gas generated (Energy Savings Trust, 2013). The Energy Saving Trust examined the impact of water consumption on water and electricity bills in the UK from 2010 (Energy Savings Trust, 2013). They found that 16% of household energy costs are water-related, or £228 of each household's energy bill. This compares with a total average water and sewerage bill of £369 amongst those surveyed. In the USA, 18% of energy bills in 2009 were accounted for by water heating (US EIA, 2013). Changing showering habits illustrate the impact of energy and water linkages for domestic bills. By shortening a daily power shower by one minute, the average user in England and Wales would cut £22 from their annual electricity bill and £26 off their water bill.

Providing this data to a consumer can be carried out either by integrating water and power consumption information from the separate meters on to a common display or by a water meter, which accepts electricity and gas use costs from the relevant utilities and adapts this to the customer's water usage.

4.3.5   Wessex Water – A Seasonal Tariff Trial

More frequent meter readings and direct data transmission to customers offer utilities the possibility of adjusting tariffs to balance supplies and demand. Summer water use, for example, is likely to be higher than during the winter, while supplies may either be similar or lower. Wessex Water carried out a smart-type meter trial (Wessex Water, 2011) on 6,000 households in 2008–2010 whereby customers (excluding the control group, who were metered but charged by rateable value) had four tariffs: standard (flat-rate), rising block (a lower rate and then higher per unit charge above a given point), simple seasonal (a higher per unit charge during the summer) or peak seasonal (a notably higher charge per unit for consumption above the winter consumption norm).

Average volume of continuous use water consumption fell from 34 litres per property per day to 15 with smart metering; for low flow (leaking lavatories and dripping taps) the fall was from 22 to 13; and for high flow (garden hose) from 12 to 2 (Wessex Water, 2011). Water consumption fell by 17% on average, with a meter, rising to 27% at peak demand periods (Wessex Water, 2011). When the seasonal tariffs were used, consumption fell by a further 6% (Wessex Water, 2012). However, the reduction in customer satisfaction resulting from seasonal tariffs was such that they are seen by the company as outweighing the extra water savings made (Wessex Water, 2012). It was also noted that some customers believed Wessex was varying its charges as a profit-boosting measure because it is a privately held company. One particular challenge was that 33% of low-income households surveyed would have to pay more under seasonal tariffs compared with 22% of other households.

4.3.6 Smart Meters and Utility Size in the USA

In America, the relatively small size of most utilities makes funding the development of a smart metering system difficult since the costs involved in collecting and monitoring the data strongly favour deployments of at least 10,000 meters. This does not affect traditional or AMR metering to the same extent.

In the USA, water utilities are fragmented and have a broad range in size. According to The Scott Report (Symmonds, 2016) 88.4 million metered accounts are managed by 10,431 utilities. This includes 95 water utilities with more than 100,000 metered accounts and an average of 243.2 thousand accounts, against 1,581 serving 10,001–100,000 metered accounts with an average of 38.2 thousand accounts and 8,755 utilities with 1,000–10,000 metered accounts with an average of 3.00 thousand accounts. Larger utilities are concerned about efficiency. Smaller utilities are more concerned about cost.

4.3.7 Sewerage Metering – What Goes In, and Out

There is a poor understanding in many urban wastewater networks as to which properties and which sewers within a property's boundary are connected to the storm sewer system, which to the foul sewer system and which are in fact combined systems whether by accident or by design. Increasing urbanisation, plus the adoption of hard standing areas within the urban landscape means that residential areas, especially where properties still have gardens are of importance in holding rainfall before it is discharged into the sewerage systems.

Sewerage metering at the property level would allow utility managers to appreciate the actual flow of water and the relationship between foul water and other waters that enters the foul water network. It would also become a tool for incentivising water recovery and harvesting.

With the exception of some commercial and industrial effluents, sewerage, whether foul (or black) sewerage or rainwater discharged at the property level is not measured. In England and Wales, for example, sewerage is still set on rateable values as set in 1989. Non-domestic properties may be assessed on their surface area to reflect rainfall runoff entering the sewerage system or by measuring the flow of effluent (as long as it is not containing solids) and its pollution loading (Wheeldon, 2015).

Sewer metering can alert customers that they are paying for a service that they do not need when it comes to rainwater drainage. For example, a customer that uses rain water or grey water for the garden, is paying for the water and sewerage but not generating the wastewater they are paying for, so there is a benefit on having their sewerage discharge monitored. This is also a benefit to the utility, as it is reducing the loading into the storm sewerage network and levelling out rainwater flow through the catchment area, as rainwater is being retained in the garden rather than being immediately discharged into the sewerage network.

Increased surface run-off due to more hard standing is affecting sewerage strategies. Either the sewerage network needs to be augmented to cope with increased peak flow during heavy rainfall (for example, the Thames Tideway Scheme (tideway.london) for diverting combined storm and foul water from being discharged into the Thames) or large-scale sustainable urban drainage system (SUDS) projects need to be implemented. The former is capital intensive (£4.2 billion in the case of the Thames Tideway Tunnel) and the latter can be slow to implement across a large, intensely urbanised area, due to planning delays and conflicts over land use. By accurately charging them for their rainwater input into the sewerage networks, customers can be incentivised to reduce the hard standing at their properties and to consider rainwater harvesting. This is particularly important where there are combined foul and storm water sewerage networks or where the interconnections between these networks are inadequately understood.

4.3.7.1 Wessex Water: Smart Wastewater Metering

Physical sewer metering would obstruct sewerage flow since anything put in its way will block due to the presence of solid materials being flushed into the sewerage network. No physical contact can take place and the technology had to be able to detect low flows in order to detect leaks and underlying water flow from outside the property.

A microwave monitor was developed by Dynamic Flow Technologies Limited (dynamicflowtech.com), after initial development at Loughborough University, with support from Wessex Water (wessexwater.co.uk) and Elster Meters (elster.com). As the properties of microwaves change between soil, air, water and pipes, these needed to be factored in. Dynamic Flow Technologies started with a series of alpha prototypes, capable of 15 times a second detection. The system has been able to detect flows at 0.02 litres per second, equivalent to 1,728 litres per day.

The meter enables the utility to compare the daily volume of water supply, wastewater discharge and rainfall for the trial property. When it rains, there is more wastewater discharged than water consumed, while in drier weather, discharge into the sewerage network will be slightly less than water consumption. When sewerage discharge is a lot less and consumption, this can be due to swimming pools (evaporation) or gardening. It may also indicate an internal leak. Permanent network flow monitoring will allow customers to understand the integrity of their rain and foul connections and to understand where these are in fact interconnected. Using real-time data, every flush can be monitored and as with smart water metering, the meter ought to be able to identify other events such as a bath being emptied, a shower and clothes and dish washing machine cycles.

For the customer, supply management signalling can encourage them to install a water butt or a soakaway to store rainwater for the garden rather than paying for it to be discharged through the sewage network.

For non-domestic properties in England, retail competition from 2017 will create a potential market, especially for customers with a significant area of hard standing. There is also the potential to develop metering in terms of the quality of the wastewater, by measuring flow and suspended solids along with the chemical oxygen demand. This would result in a 'Mogden' meter (the formula by which trade effluents are assessed for billing by utilities) which can provide data for calculating charges by effluent loading. That would allow the utility to accurately charge customers in terms of what actual wastewater and pollution loadings they generate.

Table 4.9 Examples of commercial customer leak detection and savings.

Client	Issue identified	m³/day	A$/day
Supermarket	Roof sprinkler left on	35	135
Supermarket	Faulty valve	14	40
Supermarket	Faulty valve	16	55
Supermarket	Roof cooling system leak	66	220
University	Hot water system leak	216	650
Office	Three separate leaks identified	53	195
Care home	Pipe leak	230	700

Source: Adapted from Water Group case studies.

There is the question of cost of installing sewerage meters. This would depend on the benefits especially through the potential of linking this data to the sewage network and the WWTWs so that they have flow data. For WWTWs, exceptional flows can cause major operational challenges and they are behind a significant proportion pollution incidents.

In practical terms, commercial development of smart wastewater metering is unlikely to make a significant impact before 2020–25 at the earliest, yet it could have a significant impact on overall systems operations as well as for individual customers.

4.3.8 Smart Metering and Leak Detection for Commercial Customers

Water Group (watergroup.com.au), a Sydney based company founded in 2006 offers large users a series of metering and water usage monitoring and management services with guaranteed reductions in water usage and lower utility bills. Clients include utilities, supermarkets and shopping malls, universities, offices and care homes. Table 4.9 shows examples of the daily savings in water and water bills from smart meter interventions for commercial customers in Australia.

4.4 Domestic Water

4.4.1 Domestic Devices

Domestic appliances are available for four functions; to monitor water consumption, to alert owners of water-related risks, for water harvesting and reuse, and for managing and minimising water use.

Many of the applications for water harvesting and reuse and for managing and minimising water use are not 'smart' per se. They are simply technologies and techniques that use consume less water than had previously been the case, or enable water to be beneficially reused. Their 'smart' element lies in their use being driven by the data generated by smart water (and electricity) metering. The greater the frequency of data generation and the ability for a customer to interact with it, the greater the incentive to minimise water use. The potential for Internet-linked devices creating a domestic smart network is discussed in section 4.4.7.

Table 4.10 Potential water savings for domestic appliances.

Device	Studies	Range (l/cap/day)
Washing machines	4	30–106
Shower head	3	36–46
Lavatories	3	43–111
Tap (with aerator)	2	7–35

Adapted from Percili and Jenkins (2015).

A review (see Table 4.10) of four UK based studies published between 1999 and 2011 on the potential savings from water efficient domestic devices found a broad range of identified savings compared with conventional appliances.

4.4.2 Monitoring Water Use

Domestic smart water metering at the household level was covered in sections 4.2 and 4.3. Two other aspects merit attention; the development of apps which are enabled by smart water metering and metering at the appliance and tap level. The four examples discussed below are all at the development stage.

FlowGem Limited is an early-stage developer for remote monitoring of domestic property leaks and water usage via smart phone and tablet enabled apps. The monitor (tag) is connected to the domestic pipe next to the main stop-cock. The dedicated Flowgenie app collects data from the tag and presents it in numeric and graphical ways to provide historic water flow data along with a history of leaks and water consumption for the detection of any anomalous water use. FlowGem was acquired by Centrica, the parent company of British Gas for £13 million in August 2016. Centrica is seeking to invest £500 million in domestic Internet of Things utility approaches and services by 2020. It is anticipated that FlowGem's prototype device will be developed to fit in with the rest of Centrica's domestic remote utility monitoring and management systems.

Another approach is to develop a meter that is able to break down water usage by individual devices and to generate usage data based on this. Fluid Labs (fluidwatermeter. com) is developing a FLUID smart metering unit which is attached to the main water pipe and connected to WiFi for data transmission to a smart phone or tablet. Ultrasonic measuring across the pipe allows non-physical measuring of water flow. A dedicated App allows the user to synch the meter with various household devices. Each device (lavatory, shower, washing machine and so on) can be identified through its flow signature, for example, run rate and duration, from a flushing lavatory to a washing machine cycle. This allows the meter to identify each device and when it is being used (Magee, 2015). It also alerts users to water consumption patterns triggered by burst pipes.

The AquaTrip household leakage detection system (aquatrip.com.au) is a unit fitted after the household water meter. It monitors all water flow into the property and is programmed by the user to detect anomalous water flows. Hard wire or remote control panels are also available. A valve automatically shuts off the pipe if anomalous water flow is detected. Peak/off-peak and home/away modes increase the sensitivity of the monitoring. The system can also display consumption and billing data.

Goutra (en.goutra.com) is an Algerian company which is developing taps with built in meters and meter units than can be retrofitted to taps, lavatories and showers. The devices allow for the real-time measurement water usage both for an individual instance by each device and over a period of time. Data analysis and interpretation is provided via a monitor. The intention is to match actual water consumption with 'ideal' consumption in relation to the user's circumstances. Data can be presented as numbers or graphics at the tap and household level and can be customised to highlight individual users in multi-user households. This is also a healthy reminder that smart water innovation is not in the least restricted to countries traditionally associated with engineering and IT excellence.

4.4.3 Water Harvesting and Reuse

There are a wide variety of devices and systems for harvesting rain water and for capturing and reusing grey water (bath, shower and some washing water). In both cases, water is collected by a dedicated system and it is used either for flushing lavatories or for garden watering. As both types of water are usually discharged into the household sewerage system, there will be a particular inventive to deploy one or both of these where smart sewerage metering (see section 4.3.7) is introduced.

Some elegantly simple examples of greywater recycling have been developed for the integration of lavatories and washbasins. Sanlamere's Profile 5 (sanlamere.co.uk) and Roca's W+W (roca.com) have a basin directly above the cistern, so that hand washing water goes straight to the cistern.

Another approach is to productively use the cold water that lies within hot water pipes when they have not been used for a period of time. Enviro Save Water System (Enviro Manufacturing, enviro.net.au) enables the automatic diversion of cooled water in the hot pipe network for reuse via the cold water network or a cold water storage tank. It claims to reduce household water consumption by 10%. Other examples include the Winn's Water Saver (Winn's Folly, winnswatersaver.com) which claims up to 20% household water savings and the Redwater Diverter (Redwater Australia, redwater.net.au) which claims 9.3% savings.

4.4.4 Reducing Water Consumption at the Tap Level

Domestic smart water management is of limited utility unless consumers have a suite of appliances that assist them to minimise their water usage. Such applications can be fed into a loop whereby usage is further influenced by new information generated by a smart on the effectiveness of the individual devices.

Domestic water consumption can be managed through the effective use of water efficient devices or through monitoring and manipulating water flows through devices. More specific approaches are being developed either by limiting the water flow from the individual tap or through being informed about water usage by tap.

4.4.5 Optimising Water Flow From the Tap

The Waterblade (waterblade.co) is an ABS plastic nozzle that is designed to be fitted to bathroom and cloakroom taps, primarily for offices, commercial units and the leisure sector. It was developed at the University of Brighton. A small flow of water is shaped so that it emerges as a thin sheet, resulting in a flow of 2.5–3.0 litres per minute against 10–20 litres per minute for a standard nozzle, while offering the same benefits. For a heavily used tap, for hand washing with warm water this represents a water and energy saving of up to £75 per annum. A domestic tap which runs for five minutes a day would save 13.6 m^3 per annum or £28–46 per annum (metered water bills for 2015–16 (including the wastewater charge) range from £2.05 per m^3 for Thames Water) to £3.40 per m^3 for South West Water) along with £15 in energy costs. In the UK, return on an investment of £7.50 per tap would be in two to three months depending on usage.

4.4.6 Domestic Flood Prevention

Another area, flood prevention, is becoming an early example of the application of the household Internet of Things. It is driven by the need to prevent internal flooding from burst pipes or leaks from pipes or domestic appliances. Domestic property leaks (escape of water) in the UK cost on average £2,000–4,000 to rectify, rising to £7,000 for bursts caused by frozen pipes. Escape of water claims in the UK vary between £730–912 million per annum (ABI, 2013, ABI, 2011), with claims made for 371,000 household leaks in 2010 (ABI, 2011). In 2014, domestic and commercial escape of water claims were in the region of £980 million (ABI, 2015).

Internet-based domestic data capture and control hubs have been adopted by a number of companies offering a range of dedicated services including leakage detection. The Z-Wave hub standard (z-wave.com) has been adopted by some 325 manufacturers. Water related Z-Wave enabled devices include water valves for the remote shutting of a water supply (EcoNet – econetcontrols.com; WaterCop Pro – FloodCop.com), leak sensors for leaks and floods via a separate alarm unit (Aeon Labs – aeotec.com, Everspring – everspring.com) or a dedicated app (Fibaro – fibaro.com, Fortrezz – fortrezz.com). The Insteon (insteon.com) system is based on a hub central controller ($80) that collects household data and transmits this to users via apps and a graphical interface. This allows the user to deploy devices ranging from movement detectors to remote dimmer switches. The individual water leak detectors ($35 each) are designed to be placed at point within a house where water leaks are more likely to occur, such as beside a washing machine or a lavatory unit. More units can be added to create a comprehensive network of leak detectors. The device sends a daily 'heartbeat' to indicate that it is functioning. Otherwise, it is activated when a leak is detected and sends an e-mail to the user's account.

Stand-alone systems are purely for detecting the presence of water. The WaterCop (floodcop.com) is a stand-alone leak/flood detection and alarm system with a dedicated shutoff valve. The WaterCop Pro can support 45 wireless and eight wired sensors along with repeater units for covering larger properties.

The Ark water monitoring device, developed by Ark Labs of Alabama, USA (thearklabs.com) is attached to a water heater or household water pipes, which both monitors water flow and on the detection of anomalous water usage, alerts the customer via an app to allow them to trigger a valve, therefore preventing further water losses or damage before the fault can be addressed by a plumber. The device also allows water to be remotely cut off if, for example, a tap was left on in error (Breken, 2016).

4.4.7 Water Efficient Appliances

Water saving appliances on their own offer incremental improvements; when combined, they may offer significant reductions in water consumption. A wide variety of domestic appliances are currently available or in development with water efficiency in mind. Some have been outlined above in this section. Garden irrigation systems will be discussed in Chapter 7.

Most of these devices are designed to use less water to enable a consumer to respond to data generated by a smart meter. Some also contain also smart elements. The actual effectiveness of various water 'efficient' devices has been called into question, when their limitations mean that water savings are minimised. For example, a poorly configured low flow shower will have its impact cancelled out when users compensate by taking longer showers. To be effective, water-efficient devices need to be a satisfactory 'customer experience' whether in a shower or cleaning dishes or clothes.

There are three ways of using less water in a shower: [1] through shower heads that deliver less water, but in a way which allows the user to feel that this is a normal shower; [2] by limiting the time the shower is used for and; [3] by creating a closed circuit shower system. The best example of a closed circuit system seen to date, the Orbital smart shower system, is looked at in some detail in Case Study 4.12. Kelda technology is an example of a low flow shower head and is described in section 4.4.8. Shower timers have been developed by a number of companies. For example, the Showerguard unit (Showerguard Limited, showerguard.com.au) limits a shower to a pre-set period of 2–10 minutes, with a pulse of cold water delivered as a one minute warning and no optional extension. The Shower Shortener (Davinda Innovations, davinda.com.au) offers pre-set showers of three, five or seven minutes, with a single one-minute extension.

While the emphasis for lavatories has been in low and dual flush systems, some companies have adapted a more radical approach. For example, in the Propelair (propelair. com) unit, air and water are held in a two-section cistern. Before flushing, the lid is closed to create a seal and the air pushes the content of the lavatory into the sewage pipe. This means that a flush of 1.5 litres can be used for all applications. No modifications are needed for the plumbing. For commercial applications, this means a rapid payback where the units are intensively used. Trials with WRc (wrcplc.co.uk) found 84% savings for water and 87% for energy. In commercial applications where a larger flush is needed, a unit can be added to generate an occasional larger flush when necessary.

4.4.8 Commercial and Municipal Applications

The same drivers apply for commercial and municipal consumers. Four representative examples are summarised below, covering showers, lavatories, washing machines and dishwashers. In some of these cases, domestic versions are also being developed. For schools, offices and other public buildings smart water management extends into areas such as smart water fountains, that are managed according to usage patterns, to minimise electricity consumption. When the building is closed, they go into hibernation in anticipation of re-opening.

4.4.8.1 Low-Flow Shower Heads

Kelda Technology (keldatechnology.com) offers low-flow showers for sites such as hotels, gyms and student housing. These units deliver 2.4 times as much spray force from the same water flow from a conventional shower head. Air and water are mixed in the shower head's atomisation chamber to form a spray which is projected from five nozzles. The air is injected into the chamber through a separate control and supply system that is fitted behind the shower head. The system is designed for retrofits as well as new builds. This means that a flow of four litres per minute has the effect of a conventional shower at nine litres per minute. Units have been tested and developed at Southampton University's Institute for Industry for water to air ratio in shower droplets and have related this to water flow and the 'spray momentum perception' also known as 'shower feel'. The company believes that the global shower market is worth £13 billion pa.

4.4.8.2 Vacuum Lavatories

The JETS lavatory (jetsgroup.com) was originally developed for cruise liners and aircraft. The company's Vacuumarator pump pushes air and a small amount of water through the sealed lavatory unit. The vacuum is generated by the pump with the unit being sealed through a discharge valve. The units are now marketed for trains and buses, as well as for office, residential and commercial customers, with 200,000 units having been installed since 1986.

Uninove University, Sau Paulo, Brazil had 720 units installed in 2007, reducing water consumption from 420 m^3 per day to 60 m^3 per day, saving \$1,480 a day during term time, paying for themselves in 15 months. Banco Santander's Sao Paulo office had 412 units installed, later expanded to 508, saving \$165,000 per annum. Melbourne Water's HQ in Australia had 72 units installed as part of a new build project. These units consume 0.8 litres of water per flush, equivalent to 24.4 m^3 per day less consumption than when using a conventional low flush lavatory.

4.4.8.3 Minimum Water Cleaning

Xeros (xeroscleaning.com) uses polymer beads are used as a substitute for most of the water and cleaning substances in a dedicated washing machine. The beads can be used for hundreds times, after which they are replaced. A 25 kg commercial wash will use 50 kg of beads, 1.5 million in total. The technology also enables a higher recovery rate (stained cloth than can be reused) uses lower wash temperatures and lessens fabric wear. For commercial laundry, 75–80% less water is consumed, along with 50% less energy. The company is currently developing a domestic washing machine.

4.4.8.4 Glass Washers for Caterers

Washing water is cleaned via reverse osmosis and recirculated, without rinsing being needed to limit water use to 2.0–2.5 litres per wash, compared with 9–10 litres per wash for a conventional unit. A range of programmes ensure that the wash cycle is matched to the type of glassware being cleaned. Examples include the Bracton MR/BR2 Glasswasher (Bracton Group, bracton.com) and the Winterhalter Classeg/Winterhalter UC (Winterhalter (Australia) Pty, winterhalter.com.au).

4.5   Developing Water Efficiency Standards

There have been a number of local, national and international labelling schemes for domestic and commercial water efficient appliances. Three approaches are noted. The first concentrates on approving a specific product as being water efficient (section 4.5.1 and section 4.5.3). The second sets levels of water efficiency which products can be classified within (section 4.5.2). A third sets mandatory or voluntary efficiency standards (section 4.5.4).

4.5.1   Australia – Water Efficiency Approvals

The Smart Approved WaterMark (smartwatermark.org) scheme is designed to certify water efficient appliances. It was originally developed in Australia in 2004 as one of a series of responses to a long term drought and was established in the EU in 2015. This has taken place by amalgamating national schemes into the Smart Approved WaterMark, such as the Waterwise Recommended Checkmark in the UK (waterwise.org.uk). Western Australia's Water Corporation (watercorporation.com.au) uses Smart Approved WaterMark in conjunction with its own Waterwise label under the Waterwise Products Program.

In September 2016 (smartwatermark.org/products), there were 88 approved products, 41 for domestic applications, 25 for commercial users and 22 for swimming pools. Products include soil conditioners (15), irrigation equipment (21), washing and cleaning systems and chemicals (13) and water saving appliances (6).

4.5.2   Water Efficiency Labels in Portugal, Singapore and the EU

The Certificação da Eficência Hídrica de Produtos is a national voluntary labelling scheme launched in Portugal in 2008. The main categories are outlined in Table 4.11. In addition, for a dual flush system, the low flush has to be 2.0–3.0 for A++, 3.0–4.0 for A+ to B and 3.0–4.5 for C (Benito et al., 2009).

Singapore's WELS (Water Efficiency Labelling Scheme) was introduced on a voluntary basis in 2006 and has been mandatory since 2009. It is outlined in Table 4.12. It is a progressive scheme; all applicable goods have to meet at least the 'good' criteria and since 2015, 'very good' has been the minimum standard. Good is awarded one tick on

Table 4.11 Portugal – lavatory labels (litres per main flush).

Category	Dual flush	Variable flush	Full flow
A++	4.0–4.5	–	–
A+	4.5–5.5	4.0–4.5	–
A	6.0–6.5	4.5–5.5	4.0–4.5
B	7.0–7.5	6.0–6.5	4.5–5.5
C	8.6–9.0	7.0–7.5	6.0–6.5
D	–	8.5–9.0	7.0–7.5

Adapted from Benito et al., 2009.

Table 4.12 Singapore – The WELS scheme.

Device	Good	Very good	Excellent
Showers (litres/minute)	7–9	5–7	<5
Lavatory (full flush, litres)	4.0–4.5	3.5–4.0	<3.5
Lavatory (low flush, litres)	2.5–3.0	2.5–3.0	<2.5
Washing machines (litres/kg)	12–15	9–12	<9

Adapted from pub.gov.sg/wels.

the label, very good two and excellent three. The labels inform the consumer how much water is saved by using each device. In 2011–12, 37% of washing machines sold met the three tick standard, rising to 54% in 2012–13 and 70% in 2013–14 (Benito et al., 2009 and pub.gov.sg/wels).

4.5.3 Europe's Water Label

The European Water Label's (europeanwaterlabel.eu) Water Efficiency Project was launched in 2007, changing its name to Water Label in 2009. The project has launched a suite of labels, typically giving six levels of water efficiency. There has been an emphasis on taps, showers and lavatories. In 2009 there were some 1,000 products under the label, which rose to 11,051 from 97 manufacturers by the end of 2015. Five per cent of the products are subject to external audit to maintain the integrity of the scheme (Orgill, 2015; Orgill, 2016). The three tables (Tables 4.13, 4.14 and 4.15) below are for the end of 2014.

In addition, 180 of the basin deliver less than 3.0 litres per minute. Kitchen taps are intended to fill kitchen sinks, so the flow restriction (55% delivering less than 10 litres per minute) recognises a need to lower consumption when a tap is used continually.

Water pressure varies widely across Europe (it is typically lower for example in the UK). Shower valves are usually bought for new builds, while handsets are for replacement and refurbishment. A total of 51% of shower valves have a flow of less than 13 litres per minute compared with 74% of handsets.

Table 4.13 EU Water Label: Basin and kitchen taps.

Litres per minute	Basin taps	Kitchen taps
>6.0	1,213	87
6.0–8.0	219	45
8.0–10	404	80
11–13	39	19
<13	274	123

Adapted from Orgill, 2015.

Table 4.14 EU Water Label: Shower valves and handsets.

Litres per minute	Shower valves	Shower handsets
>6.0	205	70
6.0–8.0	175	99
8.0–10	75	170
11–13	125	79
<13	542	146

Adapted from Orgill, 2015.

Table 4.15 EU Water Label: Lavatories and cisterns.

Litres per flush	Flushing cisterns	Lavatories
>3.5	24	201
3.5–4.5	368	507
4.5–5.5	161	82

Adapted from Orgill, 2015.

In the UK, lavatories are sold as a complete unit. In most of the rest of Europe, the flushing cistern is sold separately. 26% of lavatories had a flush of less than 3.5 litres, compared with 4% of flushing cisterns.

4.5.4 Voluntary and Mandatory Schemes

Table 4.16 outlines how standards have evolved over time. Of particular interest are the various European 'Eco Label' standards, which have gone through three versions between 1993 and 2010.

Table 4.16 The evolution of some water efficiency standards.

Lavatory	Date	Type	Standard
Italy (Urbino)	1997	M	5–8 litres single flush, 3–5 dual flush
United Kingdom	2001	M	6.0 litres single flush, 4.0 for dual flush
Spain (Madrid)	2006	M	6.0 litres, single flush
United Kingdom	2007	V	4.5 litres single flush, 3.0 for dual flush
Italy (Avigliana)	2007	M	6.0 litres single flush, must have dual flush
USA	2007	V	4.9 litres single flush
Europe	2015	V	4.0 litres single flush, 3.0 for dual flush

(Continued)

Table 4.16 (Continued)

Lavatory	Date	Type	Standard
Shower	Date	Type	Standard
Spain (Madrid)	2006	M	Maximum flow of 10 litres per minute
United Kingdom	2007	V	Maximum flow of 13 litres per minute
Italy (Sassari)	2008	M	Maximum flow of 10 litres per minute
Europe	2015	V	Maximum flow of 8 litres per minute
Clothes wash	Date	Type	Standard
Scandinavia	1989	V	16 litres per kg (cotton, 60°C)
Europe	1993	V	12 litres per kg (cotton, 60°C)
Europe	2005	V	9.4–19 litres per kg (cotton, 60°C) [1]
Europe	2010	V	8.0–13 litres per kg (cotton, 60°C) [1]
Dishwasher	Date	Type	Standard
Scandinavia	1989	V	1.2 litres per place setting
Europe	1993	V	1.2 litres per place setting
Europe	2005	V	0.7–1.4 litres per place setting [2]
Europe	2010	V	0.7–1.1 litres per place setting [2]

Type: M = Mandatory; V = Voluntary.
[1] Depending on the load.
[2] Depending on the number of place settings.
Schemes outlined:
United Kingdom: Water Supply (Water Fittings) Regulations, 1999; BMA Water Efficiency Labelling Scheme, 2007.
Spain: Ordenanza de Gestión y Uso Eficiente del Agua en la Ciudad de Madrid, 2006.
Italy: Ambientale al Regolamento Edilizio della Citta di Avigliana – Allegato Energetico, 2007; Variante all' Art. 8 delle Norme Tecniche di Attuazione del P.R.G., 1997; Regolamento Energetico Ambientale, 2008.
Scandinavia: The Nordic Eco-Label, 1989.
Europe: The European Eco-Label, 1993.
Sources: Benito et al., 2009, EU 2013a; EU 2013b; EU 2010a; EU 2010b.

4.6 Case Studies: The Emergence of Smart Domestic Metering and Appliances

Thirteen case studies are presented. Six consider the drivers for smart water metering, as a demand management tool at Thames Water (4.6), the preparation for smart metering in Japan (4.1) and the USA (4.8 and 4.10), the role of retail competition in Scotland and England (4.7), along with smart water metering from the perspective of an energy utility (4.3). Three look at the roll out of smart metering in Malta (4.5), Jersey (4.11) and at Southern Water (4.4). Three consider the impact of smart metering, in the context of household water use (4.2 and 4.9), and a domestic/commercial shower which both incorporates smart elements and is designed to minimise water consumption (4.12). Case Study 4.13 is a tool for enabling utilities to report information back to customers.

4.6.1   Case Study 4.1: Smart Water Metering in Japan

Japan is unusual in that its population is expected to decline significantly, from 128 million in 2010 to 85 million by 2060. This means that utilities need to plan for the future shrinkage of their systems. All domestic properties with a household connection (>97%) have water meters, which are typically read six times a year. Water utilities in Japan are organised on the city and town level, with 1,400 utilities serving more than 5,000 people. There was an AMR trial for 48,000 properties in Tokyo in 1976–98. AMRs have been widely deployed since 2000, covering 30% of the population. Since 2014, three AMI trials have been carried out. In Tokyo, this involves a visual readout for each property and e-mail alerts for anomalous water use. In Yokosua, 200 properties are being used to examine data transmission and processing. At Yokohama, AMI water and electricity meters use a common transmitter and efficiency and reliability is compared with ongoing manual readings. A further trial will start in Kobe in 2017. As most manual meters are relatively new, replacing these and the AMR units would have significant upfront costs. With universal electricity AMI metering by 2025, the aim would be to deploy AMI water meters in the wake of this, concentrating on demand forecasting, pressure management and leakage detection (JWRC, 2016).

4.6.2   Case Study 4.2: Water Use in the Home

The Energy Savings Trust has carried out two surveys on domestic water consumption in the UK. 'At home with water' in 2013 (Energy Savings Trust, 2013) was based on data which was provided online by 86,000 respondents. The second survey, 'At home with water 2', in 2013–14 (Energy Savings Trust, 2015) was a more detailed analysis of 69 households in the Thames Water area, looking at consumer behaviour. In addition, the second survey examined water flow at the household level was over a two week period with a dedicated meter that took one reading per second to evaluate the use of individual appliances.

4.6.2.1   At Home with Water

The initial survey concentrated on broad water and water-related energy consumption data. 54% of water consumed within the house is heated (baths and showers 33%, bathroom hot tap 7%, dish and clothes washers, 10% and hand-washed dishes, 4%) and 68% is consumed in bathrooms and lavatories (showers 25%, baths 8%, lavatories 22%, hot taps 7% and cold taps, 6%).

There is still a lot of room to improve customer efficiency for both metered and unmetered households and the potential for education cannot be understated. For example, 20% of metered customers let their taps run while brushing their teeth. With just 8% of people surveyed appreciating the role of water usage plays in their energy bills, there is much work needed to inform them about the linkage and its potential for reducing both water and energy usage and bills (McCombie, 2014).

A total of 49% of households surveyed used inefficient showerheads (31% standard mixer and 18% a standard power shower), with 35% using a somewhat more efficient electric mixer and 16% using efficient showerheads (5% eco power and 11% eco-mixer). There was a broad variation in shower lengths, 45% taking 1–5 minutes, 42% lasting 6–10 minutes, 12% for 10–20 minutes and 1% having showers for more than 20 minutes.

The older the lavatory the larger the flush as new standards have progressively been adopted. Six percent of lavatories date from 1940–80, 36% from 1980–2001 and 58% post-2001. A total of 41% have a dual flush mechanism, almost all being installed since 2001, and 59% have a single flush.

4.6.2.2   At Home with Water 2

The second survey (McCombie, 2015) found that consumers have a poor appreciation of the water intensity of white goods and lavatories. Washing machines are typically seen as the largest water consumers, with lavatory flushing being significantly underestimated. Unmetered households tended to disregard their water consumption, considering it to be irrelevant. For both metered and unmetered households, water ranked below energy when considering their consumption. This reflects their perception of lower bills and the lack of appreciation about how the two are connected. Water use and bills also tend to be overlooked when considering competing priorities.

The Smart Meter Advice Project provides tailored energy use advice based on real energy consumption (hot water). The two-way nature of data flow extends beyond metering information. Traditionally, utilities refrained from informing customers about their services other than to send them a bill. Customer engagement started to evolve for example in England and Wales, where the threat (which was not carried out) of water supply cuts during the 1995 drought resulted in a hostile public and political reaction, especially for Yorkshire Water, where the company's entire board of directors were obliged to resign in consequence. The ten water and sewage companies had been privatised for five years at the time, and public expectations about service delivery rose in tandem with tariffs. In contrast, there were widespread water cuts, and the use of public standpipes during the 1976 drought, but public expectations were significantly lower, with the companies at the time being state-owned.

Customer engagement has become increasingly sophisticated, moving from providing advice based upon on broad segments and house types to customising this advice based on a broad range of factors. These include households with gardens or window boxes, urban or rural location, household type (flats, mansion blocks and houses), whether the household is billed on metered consumption or a rateable value charge, household size and composition and more recently, regional differences, religious and ethnic factors and the degree of customer engagement (McCombie, 2015).

Challenges in customer engagement include perceived lifestyle compromises such as impacting their choices or pleasure. Developing equipment to fit standard-sized properties is also a problem, especially with the current trend towards smaller new builds and sub-divided properties.

With a large volume of data being generated, it should only be used if and where it can be translated into meaningful information for the utility or the customer. In addition, customers want and respond to information, rather than jargon. All communications therefore need to be scrutinised for their comprehensibility. For the customer, this means being informed about how much they spend at the household level, and from this, how much they can save. This can be taken on to consider potential savings by device. Intriguingly, people surveyed take fewer baths and showers (3.9 per person per week) than they say they do (5.4 per week).

Lowering personal water consumption is seen as a low priority because people are reluctant to compromise on personal comfort (long showers and deep baths) along with the perceived time and effort involved in using less water. Another inhibiting factor were poorly performing or broken water-saving devices, along with the feeling that such devices will not perform as they are meant to. Evidently, customer scepticism cannot be underestimated.

Other barriers included scepticism about the actual effectiveness of water-saving devices, lack of confidence in using them properly and the concern that water companies leak too much water for personal consumption to matter. Billing is also a disincentive where information is seen as being poorly and incomprehensibly presented and six-monthly bills do not inform the customer adequately about the current or potential impact of modifying their consumption.

People are interested in being informed, as long as this information means something to them. The Energy Savings Trust (EST) found that the most effective way of engaging with customers about lowering consumption was to start the process by telling them about their water consumption and its consequences. There is a willingness to accept that using less water is right, necessary and achievable, when people are properly informed about this. From there, customers can be advised about what they can do in their house without sacrificing their lifestyle. The EST's water energy calculator was found to be popular as it allows individuals to understand the water consumption of each device and to appreciate the cost impacts of both cold and hot water. This requires strong baseline information to be developed so that the customer knows where they are starting from, how they can change this and by how much and how their consumption compared with others (McCombie, 2015).

4.6.3   Case Study 4.3: Smart Metering from an Energy Utility Perspective

As smart electricity metering is well ahead of smart water metering in terms of its development and deployment, it is useful to consider its impact. A review for the American Council for Energy-Efficient Economy (ACEEE) found that the more specific and timely energy data is, the greater the savings are. Table 4.17 outlines how ACEEE report shows

Table 4.17 Energy saving compared with standard billing.

Enhanced billing	3.80%
Estimated feedback	6.80%
Daily/weekly feedback	8.40%
Real-time feedback (premises level)	9.20%
Real-time feedback (device level)	12.00%

Adapted from Ehrhardt-Martinez et al., 2010.

that a 9–12% reduction for energy can be obtained if the information is near real-time (Ehrhardt-Martinez et al., 2010).

4.6.3.1 Psychological Basis: Experiential Learning

When customers are presented with actionable information ('concrete experience') based on recent smart meter data that tells consumers where they are spending their energy (and money), this leads to observation and comprehension, then they are enabled to try and save energy and water ('active experimentation') used in heating, for hot water, and by appliances (Rentier, 2014).

In the Netherlands, Delta (delta.nl), a water and energy utility sought to see how consumer behaviour could be modified within the technological setting of the Dutch Smart Meter Requirements. Here, consumers were left to draw their own conclusions based on near realtime information and to act upon it. This was achieved without a campaign or brochures, except for a single letter to each consumer. Asked if they were able to distinguish between the bigger energy culprits from the data they were given, 48% of the consumers said that they could. This highlights the need for a suitable a publicity campaign prior to a conservation programme and if energy and water consumption and cost data is available on a suitable social media platform. This means it needs to be easily available on for example, a smart phone, tablet, laptop or PC. As with the ACEEE findings, the closer to real-time that the data is available, the greater its impact can be.

Consumers did generally appreciate the extra data, as long as it was combined in one easy graphical representation, without the need to navigate between any pages. Where data was refreshed every 60 minutes, its impact was lower in terms of 'experiential learning' since once the customer has noticed the change in energy use, the chance to act on that specific piece of information has gone. Real-time data works best when it can be appreciated in real-time.

4.6.4 Case Study 4.4: Southern Water's Smart Metering Roll-Out

Southern Water (southernwater.co.uk) adopted a comprehensive stakeholder engagement process prior to launching their smart metering roll-out. This included involving schools that had been previous water winners, promotions at local stores, events and the active co-operation of their partners; WWF (World Wide Fund for Nature); Waterwise; the Design Council; the Energy Saving Trust; Groundwork (these five entities are NGOs involved with various aspects of energy and water efficiency, with

Groundwork specialising in communicating environmental awareness to customers); Arad (smart meters); and Balfour Beatty (meter installation).

Prior to the metering roll-out, 78% of customers surveyed supported metering, primarily citing fairness as a higher priority than costs (Earl, 2016). Formal approval was granted in October 2009 and by 2015, metering increased from 40% in 2010 to 90%. The company's target is to reduce overall domestic consumption by 15 litres per capita per day 2020 (Earl, 2016) through metering and improved public awareness (Fielding-Cooke, 2014).

The universal metering project team was set up over eight months in 2009–10, based upon 40 specialist staff which operated as a directorate within the company. Southern Water found that they needed to increase their internal communications about the programme more than they had expected to and that this also involved planning the planning process and taking to account the practicalities at the customer end involved when installing meters on this scale.

For individual customers, as seen in other cases, it is essential to inform them how their bills will change from rateable value (RV) to metered and how they can bring their costs down. The costs and benefits will vary greatly with economics, age, and type of household and so on and this variability needs to be appreciated and communicated at the outset and use the smart data to help customers get to grips with their water (and energy) consumption.

Water efficiency audits are being now being carried out to help vulnerable people, especially households where consumption is at least 20% above average and the bill accounts for more than 3% of household income. 28,000 audits are planned between 2015 and 2020 and target customers will also have had a smart meter and a bill designed to reflect their water use, along with their using a combi boiler and an electric shower. This is part of a strategy to avoid future bad debt of £147 per customer for the most vulnerable (Earl, 2016).

The initial roll-out in 2010 covered 350,000 new meters with 60,000 extant meters exchanged. In addition, there had been 30,000 Green Doctor customer water efficiency visits since 2010. The message, 'Save water, save energy, save money' or saving water saves you energy and saves you money was broadcast in three phases; eight weeks before installation at a property, four weeks before and on the day of installation. The programme has become a significant opportunity to engage with customers and other stakeholders on both smart metering and their other activities in general and using this to improve the transparency of their customer communications. That is a beneficial effect of an unprecedented degree of face to face customer engagement, rather than along the traditional channels.

Customer affordability is a problem which is brought to the fore when universal metering is introduced. To date, 60% of households connected have seen their bills fall by an average of £12 per month, while 40% have seen their bills rise by £14 per month. Further work is needed in helping the latter group to reduce their water consumption, along with targeted schemes for supporting low income households. Southern Water also used IncomeMAX when contacting customers about tax credits and unclaimed benefits since 2010 and this scheme has secured £2 million in payouts to eligible customers.

How effective has smart metering been at Southern Water? Ornaghi and Tonin (2015) surveyed water consumption amongst 54,664 customers from pre-meter installation to their fifth six-monthly bill after installation. They found that consumption fell by 12.5% before the first bill, which is the 'anticipation effect' and from 1.5 years, it fell by 16.5%, where it remained. In essence, smart metering technology is seen as 'amazing', but in reality it is only as good as the user and how the user is informed and motivated to use it.

4.6.5 Case Study 4.5: Malta's Smart Water Metering Roll-Out

The Water Services Corporation of Malta (WSC, wsc.com.mt) smart water and electricity metering programme is also covered in some detail in Chapter 5. This case study concentrates on its experience with smart water metering. There were two AMR pilot projects in 2003–06, but in 2009, the WSC decided to adopt a fully smart AMI approach and to integrate this with the island's other utilities. In addition, it was decided to replace all meters that were over 10 years old. A total of 80% of customers have supported the smart metering programme, and the WSC aims to win over the other 20% in time. Public support is necessary for the installation of the radio frequency transmitters as well as for the meters (Pace, 2014).

Leakage was identified and quantified by comparing the actual 2–4 am water flow with the assumed water flow (legitimate night consumption). Here, smart metering is allowing utilities to differentiate between small and large customer leaks. Data profiles are used in order to understand where unusual customer behaviour is taking place. A one-month domestic smart meter reading in August to September 2012 shows in a dramatic manner the potential for savings. Out of a total consumption recorded through the meter of 29,537 litres, leakage was 19,344 litres against useful consumption of 10,193 litres; 65.5% of the water billed for was in fact wasted.

Using traditional meters, the reading process cost €1.2 million pa. The smart system has annual field maintenance costs of €420,000, with an annual licence fee of €80,000, along with €50,000 on upgrades and depreciation of €1,200,000, giving an annualised cost of €1.73 million. Traditional meters also have maintenance and depreciation costs, which while not given, highlight the benefits from not having to read meters manually. Tangible benefits of €1.6 million per annum were seen, including improved customer service, less customer disturbance (no meter reading visits) and fewer billing disputes, improved consumer profiling, more effective detection of tampered meters and internal leakage detection. Improved cash-flow through a two-month billing cycle and the elimination of estimated customer bills is improving cash-flow by more than €1 million per annum. Case Study 5.6 considers Malta's smart water programme from the utility perspective.

4.6.6 Case Study 4.6: Smart Metering and Demand Management for Thames Water

4.6.6.1 The Need for Metering

Thames Water (thameswater.co.uk) serves parts of London and South East England, a region where over 40% of renewable water resources are already being extracted. At the time of Thames Water's privatisation in 1989, the population of Greater London had fallen from 8.61 million in 1939 to 6.70 million in 1988. In 2011, it was at 8.63 million, its highest ever (GLA, 2015) and projections for 2050 range from 9.5 million to 13.4 million, with a median of 11.3 million (Tucker, 2014). Meanwhile, the population of South East England rose by 8% between 2001 and 2011, to 8.6 million (ONS, 2012).

At the same time, customer usage has increased by 30% since the 1980s. As a result, Thames Water forecasts a dry year demand shortage of 133,000 m^3 per day by 2020 and 414,000 m^3 per day by 2040.

Much of the distribution network is in poor condition and the utility currently faces problems regarding poor data quality along with inflexible billing and IT systems. A move to smart metering is seen as a central element in addressing these shortfalls. In 2015, 24% of London households served by the company had a traditional water meter along with 44% in the Thames Valley; or a 30% overall penetration rate. Traditional metered customers use 12% less water than non-metered customers (Baker, 2016).

4.6.6.2 Deploying the Meters

For Thames, the smart metering programme is seen as an opportunity to renew that way their customers perceive them and to be able to engage with them on the importance of water efficiency. The company plans to have at least 80% meter penetration by 2030, through a progressive roll-out. Legal powers were gained in 2012 for compulsory meter installation. The installation of 1,457 Arquiva and 2,551 Sensus Homerider smart meters were carried out in 2011 with trials running from 2012–15 in five DMAs, two each in London and Reading and one in Swindon. These trials found that a properly installed and managed fixed network system can deliver more than 95% data collection success (95% for Arqiva and 80% for Homerider, both of which are improving with time) via Temetra meter readers.

The formal compulsory meter roll-out programme started in 2014, starting with 4,100 meters in Bexley, London. At the start of 2015, there were approximately 300,000 AMR meters installed at Thames. In 2015–16, 40,000 meters were installed and the target for 2015–20 is for 441,000 meters, and 1 million by 2030. Including optants, replacements and new connections, a total of 3 million smart meters will be installed by 2030.

4.6.6.3 Findings from Fixed Network Trials: 2012–15

A trial with 826 customers found that consumption fell with increased metering frequency, from 186 l/cap/day with readings every 60 minutes (n = 398) to 138 l/cap/day when readings were taken every 15 minutes (n = 123). Likewise, consumption also falls with time from adoption, from 177 l/cap/day after seven days (n = 315) to 162 l/cap/day after 30 days (n = 238).

The generation of regular data proved to be effective at identifying customer side leakage (CSL), as would have been expected. Information about such leaks is triggered when there a continuous flow of more than 25 litres per hour for 14 days from 60-minute data. This was a particular boon with inactive accounts. At the same time, actual customer water usage was found to be higher than expected, suggesting that the prior estimates for customer side leakage were too high (Baker, 2016).

4.6.6.4 Preparing for the Migration from AMR to AMI

The network is designed to initially support AMR enabled meters with electronic registers and a fixed network data capture system, with common standards throughout the network. It is based on local communication equipment, which is paired to the meter and automatically switches to fixed network mode when a radio signal is detected. Radio signals are sent from masts that cover over 97% of the meters. Data includes notifications for defects management, fault reporting and KPI (utility Key Performance Indicators set by Ofwat) performance reporting.

In 2015, the AMR network was mainly being used for meter reading and billing, with limited use of other capabilities being made at the time outside specific trials (Hall, 2015). The system is capable of being upgraded to full AMI without any subsequent site visits. The contracts with Arqiva and Sensus are for 15 years, with five-year break clauses. The meters have a guaranteed operating life for at least the duration of the contract, thereby minimising the need for further site visits (Baker, 2016). During the trials, meter readings were taken every 15 minutes and the data uploaded every four hours. The company is preparing to receive 35 billion reads per annum from 3 million smart meters when the roll-out and AMI migration is complete. On average, this works out as one reading per 40 minutes (Baker, 2016).

4.6.6.5 Customer Engagement and Awareness

Informing and involving customers is an essential part of any compulsory introduction programme that directly affects them. Thames addressed this through providing customers with an awareness and survey appointment. All customer contact is carried out under a common brand with a dedicated team of staff. Early engagement with all identified stakeholder groups and the local media is essential in order to build confidence in the forthcoming programme. This also means that all communications material with customers needs to be tested prior to each formal roll-out. It was evident that any utility contemplating a compulsory smart water meter roll-out programme should not underestimate the cost and complexity of delivering this in a manner that maintains and enhances customer goodwill. It is also essential for water utilities to fully appreciate that smart domestic metering started with the energy industry and they are still far ahead in most aspects of this service.

Given the importance water holds for many faiths, it is likewise necessary to tailor customer interactions to reflect their particular faith with specific messages being developed and disseminated during the trial period for developing a more comprehensive approach during the main roll-out.

Thames borrowed from the experience of the television digital switchover campaign in the UK, starting with a trial at one transmitter in 2005 and was rolled out between nationally 2007 and 2012. Initial contact is through a letter and explanatory leaflet, followed up by a customer visit. During the initial Smarter Home Visits trial period in 2014–15, Thames completed 6,046 Smarter Home Visits, with a typical visit taking 30–45 minutes and installed 15,903 water and energy saving devices. In consequence, with the installed meters, they found that these visits resulted in a combined saving of 446,956 litres of water per day (163,139 m^3 per annum), equivalent to 73.9 litres per household per day (27 m^3 per annum) or 28.1 litres per device per day (10 m^3 per annum). This is equivalent to an annual saving of £55 for metered water bills, and £50 for energy bills, through lower hot water use.

In the extended trial to 2016, approximately 70% of households contacted took up the Smarter Home Visit offer. This included 30,000 Smarter Home Visits in the London area, under the Progressive Metering Programme, along with 7,000 Smarter Home Visits in the Thames Valley area. A total of 86% of customers visited recalled receiving information about metering and 80% were familiar with the benefits of metering, suggesting a broad base of customer acceptance can be developed for a compulsory programme. By 2017, 60,000 visits had generated savings of 2.5 million litres per day, or 42 litres per property per day (Brockett, 2017).

When the meter is installed, customers receive a 'Your new meter' leaflet, and a survey card and accessed household card are completed. After the meter enters service, the customer receives an activation letter and a letter and a letter outlining costs and benefits of smart meter-based billing. This is in turn followed by the Smarter Home Visit, to inform customers about the water saving devices available.

As part of the preparatory process, once the meter is activated, customers can access a virtual metered account to compare with the standard non-metered account and that they can opt to switch to the metered account early. Customer access their data via TW's portal and a secure connection. A customer dashboard records daily usage by month and puts it into context of costs and savings. Overall, a two-year customer transition period after the installation needs to be factored in.

Water efficiency 'TAP App' apps are offered, providing interactive advice about leakage, devices and general information on water consumption. TAP App outputs can also be device-specific. For example, for a washing machine, by inputting how often it is used and at what temperature, the customer can be informed about how much this device costs for water and for energy cost and how this could be improved. TAP App water savings reports can also be e-mailed or posted to customers. The TAP App also reports to Thames, outlining customer progress and the scope for further savings.

Customer data inputs include the type of heating, number of people in the property and the type of property. Outputs can be broken down to their water and energy cost and water, energy and carbon impact by room and device and how they can be made more efficient. Water usage is compared with the national average and a potential target based on identified savings.

4.6.6.6 Benefits Identified

Benefits identified to date include an improved perception of the utility as a brand and its image through the effect of improved data quality and more relevant tariffs. Demand management is affected on a series of levels, including identifying and managing customer side leakage; engaging customers in demand management through customer consumption viewers; and better appreciating water flow through the network. Active leakage control also reduces demand and mains replacement programmes can start to be related to actual needs rather than assumed network quality.

Customer side and distribution network leakage are both expected to be progressively reduced as the quality of information received improves through increased data availability and wider adoption as network coverage increases. Improved access to customer water consumption data allows further and more specific demand management initiatives to be developed, for example, the 'Leaky Loo' fixed network trial. This level of detail and speed of reception can only be obtained through smart metering and it was found that by repairing leaking lavatories, an average of 405 litres per day ($148 \, m^3$ per annum) was saved, roughly equivalent to an average household's normal consumption.

Customers gain from improved and more accurate billing, along with an improved service based upon the network's actual, rather than assumed performance and the ability to react to failings before the customer is necessarily aware of them. Removing the cost of customer side leakage from bills is almost certain to be supported by customers.

Tariffs can be developed that more accurately reflect customer concerns, thereby improving the willingness to pay and to modify water usage. Customer interaction in turn will be improved through being able to interact with their water use data via a 'My Meter' web page and the longer term potential to customise the data they receive. In one extreme case, during the trial period, two leaks were found at a customer's property in south east London that were losing 50,000 litres per day (Brockett, 2017).

4.6.6.7 Risks to Consider

There are risks associated with a compulsory metering programme, chiefly that it highlights potential affordability concerns, it inevitably removes customer choice, and that such a radical move means that there is little room for perceived errors. For a water utility, it can only take one mistake to tarnish the customer's experience.

Thames has seen 'some support' from Ofwat, noting that their approach appears to be 'here are the rules, now work within them' rather than specific support from the regulator in practical terms.

4.6.6.8 Going Forward

Taking forward customer engagement and performance improvements will look at how savings in water usage can be maintained and enhanced; how much customer engagement is needed; and what information provided to customers and incentives offered work most effectively. Another continuing priority will be desegregating the customer side reductions in water consumption into actively adopted water efficiency measures, identifying and dealing with customer side leakage and the simple effect of having a new meter.

4.6.7 Case Study 4.7: Retail Competition in England and Scotland

Scottish Water has been open to retail, competition for 152,000 non-domestic customers since 2008. Scottish Water, the state-owned utility provides water services in this segment via Business Stream. Its market share was 98% in 2013 when there were four companies competing for the market, and 95% four months after eight new competitors entered the market in June 2013 (WICS, 2013), falling to 75% by 2015. In 2015–16, there were 23 companies offering retail services (Scottish Water, 2016).

A dual customer retention strategy has been developed, using high-frequency, low-depth interactions such as advertising (maintain customer awareness) and the company website. Medium-frequency, medium-depth interactions involve assisting customers regarding their infrastructure contact and low-frequency, high-depth interactions include customer service calls to ensure satisfaction about prices and service levels, along with individual visits. Customer contact did not exist prior to 2008, and the main concern now is to maintain the company's presence and appreciation amongst its extant customer base (Wallace, 2015).

In England, 1.2 million non-domestic water and/or sewerage customers will be allowed to choose their retail service provider from April 2017. In Wales, this is limited to customers who consume more than 50 megalitres (Ml) of water per year. In the case of Scotland, two of the chief drivers have been customer savings through improved efficiency and better customer service. Smart metering has been offered as an enabling tool in both cases. Ofwat, the England and Wales water regulator is also looking to open the domestic market to retail competition. If this happens, it will not take place before 2020–25 at the earliest.

4.6.8 Case Study 4.8: Preparing for a Smart Meter Roll-Out in the USA

Round Rock, Texas has 110,000 people in 35,000 households in an area of 26.3 square miles. The utility decided to upgrade from AMR meters to AMI in order to improved customer service, lower NRW and increase the efficiency of their operations (Zur, 2015).

Upgrading to a Master Meter (mastermeter.com) Allegro AMI network was installed along with Harmony MDM (meter data management) software was carried out over one year. The two-way AMI system provides 24 -hourly readings along with service interruption alerts.

The base station has a range of 3–4 miles and can handle up to 75,000 units. Each base station is linked to 5–7 repeaters with a range of 7–8 miles, capable of reading from 1,000 meter units. On completion, there was a 99.6% reception rate. A total of 200 meters were installed each day by the installation team.

Implementation was rolled out in six phases:

1) Ensure business continuity, especially with regards to customer billing.
2) Recruit the personnel needed, including for IT, billing, water resources, customer support and the field team.
3) Adapt the product to ensure it meets your exact requirements. For example, alerts, reports and dashboards.
4) Staff training.
5) Delivering immediate value to the customer and the utility. This includes a phone app and an internet portal, along with delivering simple alerts such as home leaks and water theft for customers, and managing the field teams.
6) Creating long-term value. This is primarily for the utility, including developing an automated monitoring and management process, letters and alerts for customers, work orders management, addressing non-revenue water through leaks detection and DMA and a full system analytics module.

4.6.9 Case Study 4.9: Reducing Water Consumption in Melbourne

Three utilities provide water to 4.2 million people in Melbourne, Australia in 2012; 1.6 million domestic and 153,000 business customers. Per capita water consumption fell from 247 l/c/d in 2000–01 to 208 in 2005–06, 166 in 2007–08 and 147 in 2010–11, rising marginally to 149 in 2011–12 (Gan and Redhead, 2013).

The reduction in water use reflected a number of demand management initiatives. By 2012, 72% of showers had a flow rate below 8.0 l/minute, 89% of lavatories being dual flush with 60% having a maximum flush of 6.0 litres or less. A total of 36% of washing machines had at least a four star rating, along with 19% of dishwashers. Water consumption was 26% higher in summer reflecting the impact of garden watering and evaporative coolers. Water usage increases by 94 litres per household per additional household member, meaning that the larger the household the lower the per capita usage, falling from 240 l/c/day for a one person household to 120 l/c/d for households of five or more.

4.6.10 Case Study 4.10: Smart Meters in the USA, A Utility Perspective

Polling amongst the US water utility companies that attended the Smart Grid Summit's Smart Water Summits in 2014 and 2015 give some insights about utility attitudes towards smart metering. It is to be assumed that companies that send delegates to these events are more likely to be interested in smart water than companies that did not attend (Zpryme, 2014; Zpryme, 2015).

In 2014, 49% of utilities represented used AMR and 24% AMI, with 24% using neither. A total of 42% stated that they were planning to use AMR and 58% for AMI, with 12% stating that they did not plan to use AMR and 20% not intending to use AMI. Most planned AMR deployments are for the medium-term, 89% not being deployed within the next two years; 52% of planned deployment was within the next year and the rest later on (Zpryme, 2014).

In contrast, there was appreciable uncertainty about meter data management systems in the 2014 survey; 31% of respondents had a system in use, 9% had one partially installed and 3% planned to install one. A total of 38% did not expect to install a meter data management system and 19% did not know about meter data management systems (Zpryme, 2014).

At present, smart metering is seen as a utility tool rather than one for informing customers. When it comes to offering customers the ability to manage their water usage, in 2014 this was already the case for residential customers at 3% of utilities and 1% for commercial customers. This is planned for 40% of residential and 39% of commercial customers and not planned for 57% of residential and 60% of commercial customers (Zpryme, 2014).

Concerns about smart metering in 2014 (Zpryme, 2014) were principally about cost (78% of utilities), with data collection, communications systems, billing, IT support and the smart meter network noted by 26–36% of utilities. Other concerns were customer acceptance (19%) and the lack of skilled staff (14%).

4.6.11 Case Study 4.11: Jersey Water, Using AMR and AMI

Jersey Water supplies 19 Ml/day of water to 100,000 people via 38,000 connections with a 580 km of network. The utility has 120 days of water storage capacity and is dependent on surface water resources. At the same time, the population is rising and the utility faced adopting new supply approaches (desalination) or demand management (Smith, 2015).

Metering has been developed on a gradual basis. In 2003, all new connections had to have a meter, with overall adoption below 10%. By 2009, a switch to metering for all changes of occupier saw metering rise to 30%. In 2010, a universal metering plan was adopted, starting at 36% in 2010 and rising to 84% by 2015 with the aim of universal coverage in 2016/17.

The network being adopted currently uses 80% radio enabled meters and 20% encoded meters (walk-by AMR). The system has been designed to accept more advanced demand, customer and network management approaches when these are required.

4.6.12 Case Study 4.12: Orbital Systems – A Water Efficient Power Shower

The Orbital Systems (orbital-systems.com) shower was developed by Mehrdad Mahdjoubi based on a previous project for water efficient systems for a Mars mission by

NASA. The shower system reuses five litres of water in a closed loop, where the water if filtered and reheated as long as the system is turned on. The company is based in Malmo, Sweden and started public trials in Malmo during 2013. Commercial sales to leisure and healthcare operators began in December 2014 and in March 2015, $5 million of early stage funding was raised.

A typical shower would use 15 litres of water against 150 for a ten minute shower (the norm in Sweden) at 15 litres per minute. The shower delivers 15–22 litres of water a minute, comparable with a power shower. Five litres of water are consumed to start the system, and there is further wastage from splashing and water taken up by the filters. In a typical application, the company claims it uses 90% less water and 81% less energy than a conventional shower generating the same amount of water.

The micro capsule removes larger contaminants such as sand, skin and dust, costs €20 and will treat 15,000 litres of water (€0.02 per shower). A nano capsule treats 50,000 litres of water for viruses, bacteria and contaminants and costs €80 to replace (€0.04 per shower).

The shower has three smart elements. The quality of the shower water is monitored in the base unit, to determine if it needs to be treated and when to flush it away at the end of the shower. LEDs on the floor of the shower unit warn when the capsules are due for renewal. A dedicated app allows shower use data (and savings) to be monitored, along with treatment capsule status.

According to Orbital, a domestic unit in London (Thames Water) providing four seven-minute showers each day would save 146 m^3 of water (at €3.18 per m^3) and 4.85 mWh of electricity (at €210 per mWh) each year, lowering the combined utility bills by €1,089 per annum. Cost savings depend on the customer's cost of water and electricity as well as the number, temperature and length of showers used. Due to the high cost of water in Copenhagen, its first international order was to a public baths in the city in December 2015.

A floor unit, for new bathrooms costs €4,295 and a retrofittable cabin costs €5,295. The company aims to reduce this to approximately €2,800 by 2019 as it goes into volume production. A domestic unit in Copenhagen would pay back the unit's cost and filter replacements in 17–21 months (according to the unit chosen), compared with 41–52 months in London, where utility bills are appreciably lower (Hickey, 2016).

4.6.13   Case Study 4.13: Enabling Utilities to Communicate Meter Readings

Droupcountr (dropcountr.com) was originally developed in 2013 and became a formal project in 2014. The service was launched in Folsom, California in September 2014 as a one-year trial which was subsequently extended to a three-year contract. Dropcountr uses a utility administration dashboard, which analyses the raw usage data and enables utilities monitor water usage in realtime. It can be used with both smart and traditional meters. The customer's water budget is calculated through data on each property including irrigation practices and use of water appliances and their overall water usage is presented in the context of billing tiers. Leak alerts are also provided where smart meters are used.

Customers include the cities of Tustin, Rialto, Loma Linda, Fullerton, Austin Water (Texas) and Liberty Utilities, an investor-owned utility; a total of 10 utilities with 0.5 million customer accounts. In Folsom, California, consumption is 8% lower with

Dropcountr users compared to non-Dropcountr users, rising to 12% for higher water users.

Customers in an area where Dropcountr is not present can use a 'utility poke' app that geo-locates the customer and inform the utility that the customer is interested in using the service. The service identifies high users, or which addresses are engaged in rebate programmes throughout their district. The longer term aim is to develop water budgets at the address level rather than expecting a standard usage reduction.

In September 2015, OmniEarth (onmiearth.net) formed a partnership with Dropcountr to provide the Santa Ana Watershed Project Authority (SAWPA) with digital water conservation information for its customers which is being funded through California's Emergency Drought Grant programme. OmniEarth is analysing customer water consumption to identify those customers with the greatest potential to conserve water. It will deliver individualised water conservation recommendations directly to customers via Dropcountr's mobile technology with water usage advice developed for the customer. This will also reduce the amount spent to monitor progress towards conservation goals to reduce water consumption.

Monthly meter readings can be used, albeit the service is more effective with more rapid readings, ideally via AMI; 20% of their customers were using an AMI in 2016. Revenues are generated from the utility and the apps are free for end users to download via Android or iPhone. The platform allows utility staff to filter their accounts into groups for targeted communications. Customer contact depends on the frequency of meter reading and these are translated into consumption and price, along with peer-comparison. For the latter, comparisons are only made for similar properties. The service also includes water conservation advice, including indoor and outdoor water use rebates (Lohan, 2016).

Conclusions

This chapter has considered at some length domestic water metering, smart water metering and its impact on domestic demand management, especially for the development and refinement of water efficient domestic appliances. As noted in Chapter 3, a meter is not necessarily part of the smart water network. Rather it is an information gathering device and the 'smarter' the meter is, the more information can be obtained from each device. Against this is the perception that the smart meter is the means by which almost all utility customers will experience a smart network in practice, and its primary role as a driver for influencing consumer behaviour, especially through demand management. A network where there is both smart domestic water and sewer metering offers even more information, especially regarding consumption within the property and the interaction between rainwater and piped water at the household level. Smart sewage metering is likely to be rolled out slowly and selectively at the domestic level, but it is a potentially powerful source of water and wastewater data generation.

Smart metering requires appropriate tariffs in order to incentivise consumers to change their water consumption behaviour; the higher the tariffs, the greater the incentive. This reaches its conclusion in Denmark, where Copenhagen has the world's highest tariffs. As a result, high efficiency consumer goods such as internally recycling showers (see Case Study 4.12) can be developed despite their high cost.

AMI does depend on collaboration between various parts of each utility to ensure that there is a coherent relationship with customers in offering and explaining smart metering, especially about how smart metering works in reality. It is particularly important to ensure customers appreciate that smart water billing does not necessarily involve added complexity. Such complexity is indeed available where the customer seeks more detailed information and that data can be to some extent customised to meet their interests. In turn, this means that customer service has to be more responsive, using the improved flow of information to assist customers and to show them how this can be a benefit. This entails moving from a passive to a more active form of customer engagement and using the improved communication as a platform for building customer trust and indeed support. Customer relationships may well evolve, and the way that they will change will not be apparent until both the utility and its customers more fully appreciate what smart metering offers them, so utilities need to be ready to modify their approach to customers as circumstances change.

Domestic water and sewage metering are two aspects of smart water metering. In Chapter 3, district metering was mentioned, and this will be looked at in greater detail in Chapter 5. It is a tool for understanding and identifying local water loss within the district metering area and for managing water pressure management for optimising leakage rates within each area.

The impact of household water meters also reflects how tariff structures will evolve. Seasonal or even day/night tariffs would influence patterns of water usage, encouraging for example more selective watering of gardens in summer (to be discussed in Chapter 7) at one extreme and the smoothing of daily water demand across the 24-hour cycle at the other. As well as overall household water loss, internal leaks can be detected using devices within the house by developing water monitoring within a Home Area Network. All these measures also result in more detailed data loops to further optimise water consumption.

At the other end, foul water meters are used in mapping urban foul water networks, along with urban storm water metering for urban storm water network mapping, ensuring flood resilience and as a potential warning for consumers about sewer loading.

References

ABI (2011) UK swimming in household leaks. Association of British Insurers, London, UK.

ABI (2013) Burst and frozen pipes (escape of water). Association of British Insurers, London, UK.

ABI (2015) Key Facts 2015. Association of British Insurers, London, UK.

A&N Technical Services (2005) BMP Costs and Savings Study. A Guide to Data and Methods for Cost-Effectiveness Analysis of Urban Water Conservation Best Management Practices. A&N Technical Services Inc., Encinitas, USA.

Baker S (2016) Thames Water Smart Metering Programme. Potable Water Networks: Smart Networks, CIWEM, 25th February 2016, London, UK.

BCU (2015) Case Study for Metering and Sub-metering. Brunswick County Utilities.

Beal C D and Flynn J (2014) The 2014 Review of Smart Metering and Intelligent Water Networks in Australia and New Zealand. Report prepared for WSAA by the Smart Water Research Centre, Griffith University, Australia.

Benito P, et al. (2009) Water Efficiency Standards. Bio Intelligence Services and Cranfield University, Report for European Commission (DG Environment), 2009.

Bentham D (2015) SMART Water: The UK Business Case. SMi Smart Water Systems Conference, London, April 29–30th 2015.

Bookwalter G (2016) Naperville water meters too costly to read remotely, director says. Naperville Sun, 24th February 2016.

Boyle T, et al. (2013) Intelligent Metering for Urban Water: A Review. Water 5: 1052–1081; doi:10.3390/w5031052.

Breken T (2016) Ark Labs hopes to make a difference with new water conservation device. Tech Alabama, 3rd August 2016.

Carey P E (2015) AMR/AMI Feasibility Study. MC Engineering, Orangevale, USA.

Carey P E (2014) City of Orland. Meter and Water Loss Management System Report. MC Engineering, Orangevale, USA.

CIWEM (2016) Water efficiency: helping customers to use less water in their homes. CIWEM, London, UK.

DS&A (2016) Advanced Meter Infrastructure (AMI) Feasibility Study. Report to Orange Water and Sewer Authority. Don Schlenger and Associates, LLC, New York, USA.

EA/NRW (2013) Water stressed areas, final classification. Environment Agency, Bristol, UK, Natural Resources Wales, Cardiff, UK.

Engineer S (2015) Water efficiency past and present. Presentation at Water Efficiency, Past, Present, Future, Waterwise Conference, London, 19th March 2015.

Earl B (2016) Smart Water Efficiency and Affordability. Presentation at Accelerating SMART Water, SWAN Forum, London, 5th–6th April 2016.

EBMUD (2014) Advanced Metering Infrastructure (AMI) Pilot Studies Update, East Bay Municipal Utility District, Finance – Administration Committee, East Bay, USA.

Ehrhardt-Martinez K, Donnelly K A and Laitner J A (2010) Advanced meter reading initiatives and residential feedback programs: A Meta-Review for Household Electricity-saving Opportunities. The American Council for Energy-Efficient Economy, Washington DC, USA.

Elster (2010) Multi-Jet hybrid supports water conservation objectives in Saudi Arabia. Elster Meter, Mainz-Kastel, Germany.

Energy Savings Trust (2013) At home with water. Energy Savings Trust, London, UK.

Energy Savings Trust (2015) At home with water 2, Energy Savings Trust, London, UK.

EU (2010a) Commission Regulation (EU) No 1015/2010 of 10 November 2010 implementing Directive 2009/125/EC of the European Parliament and of the Council with regard to ecodesign requirements for household washing machines.

EU (2010b) Commission Regulation (EU) No 1016/2010 of 10 November 2010 implementing Directive 2009/125/EC of the European Parliament and of the Council with regard to ecodesign requirements for household dishwashers.

EU (2013a) Commission Decision of 21 May 2013 establishing the ecological criteria for the award of the EU Ecolabel for sanitary tapware (2013/250/EU).

EU (2013b) Commission Decision of 7 November 2013 establishing the ecological criteria for the award of the EU Ecolabel for flushing toilets and urinals (2013/641/EU).

Fielding-Cooke J (2014) Inside looking out. SMi, Smart Water Systems Conference, London, 28–29th April 2014.

Gan K and Redhead M (2013) Melbourne Residential Water Use Studies. Smart Water Fund, Melbourne, Australia.

GLA (2015) Population Growth in London, 1939–2015. GLA Intelligence, Greater London Authority, London.

Godley A, Ashton V, Brown J and Saddique S (2008) The costs and benefits of moving to full water metering. Science Report – SC070016/SR1 (WP2) Environment Agency, Bristol, UK.

Halifax Water (2014) AMI Technology Assessment and Feasibility Study. Halifax Regional Water Commission, Halifax, Canada.

Hall M (2015) What is the transition from AMR to AMI? SWAN Forum 2015 Smart Water: The time is now! London, 29–30th April 2015.

Hall M (2014) Pioneering Smart Water in the UK. SMI Smart Water Systems Conference, London, April 28–29th 2014.

Hamblen M (2016) Cedar Hill, Texas, relies on wireless meters and customer software. Computerworld, 21 July, 2016.

Hickey S (2016) The innovators: looped water system for Earth friendly shower. Guardian Sustainable Business, 21 February 2016.

Honeywell (2013) Water Meter AMI Project proposal for the City of Port Townsend. Honeywell Building Solutions.

Hooper B (2015) 20 years of water efficiency. Presentation at Water Efficiency, Past, Present, Future, Waterwise Conference, London, 19th March 2015.

Iagua (2010) Informe AEAS sobre 'Tarifas y Precios del Agua en España'.

IBM (2011) Dubuque, Iowa and IBM Combine Analytics, Cloud Computing and Community Engagement to Conserve Water. IBM, New York, USA.

IB-Net (www.ib-net.org) accessed April 2016.

Javey S (2016) Managing water use demand with feedback. Presentation at Accelerating SMART Water, SWAN Forum, London, 5th–6th April 2016.

JWRC (2016) Smart Water Metering in Japan. Japan Water Research Centre, Tokyo, Japan.

Kelly F (2016) Roll-out of water metering under review. Irish Times, 14th May, 2016.

Lohan T (2016) An App That Helps You Save Water and Money. Water Deeply, 24th March 2016.

Lovell A (2016) Customers and Tariffs. Presentation to Water Efficiency: Engaging People, Waterwise Annual Efficiency Conference, London, 2nd March 2016.

Magee C (2015) FLUID Is A Smart Water Meter For Your Home. Tech Crunch, 15th September 2015.

McCombie D (2014) At home with water. SMi, Smart Water Systems Conference, London, 28–29th April 2014.

McCombie D (2015) Thinking outside the box – smart water links to energy. SMi Smart Water Systems Conference, London, April 29–30th 2015.

Miller E (2015) Water Wiser: New smart water meters aimed at spotting leaks before bills climb. Santa Fe Reporter, 1st July 2015.

MandSEI (2011) City of Ottawa launches water AMI program. Metering and Smart Energy International, 12 April 2011.

MandSEI (2016a) US city agrees to roll-out of 34,000 units. Metering and Smart Energy International, 4 February 2016.

MandSEI (2016b) Canadian utility secures regulatory approval for AMI roll-out. Metering and Smart Energy International, 11 October 2016.

NAO (2007) Ofwat – Meeting the demand for water. National Audit Office, London, UK.

Nussbaum D (2015) Why Water Efficiency Matters. Presentation at Water Efficiency, Past, Present, Future, Waterwise Conference, London, 19th March 2015.

OECD (1999) The Price of Water: Trends in OECD Countries, OECD, Paris, France.

OECD (2007) Financing water supply and sanitation in ECCA Countries and progress in achieving the water-related MDGs. OECD, Paris, France.

OECD (2012) Policies to support smart water systems. OECD, Paris, France.

Ofwat (2000) June Return for 2000. Ofwat, Birmingham, UK.

Ofwat (2005) June Return for 2005. Ofwat, Birmingham, UK.

Ofwat (2010) June Return for 2010. Ofwat, Birmingham, UK.

Orgill Y (2016) European Water Label Annual Review, Water Label, Keele, UK.

Orgill Y (2015) European Water Label, Scheme, Roadmap and Vision 2015, Annexes A–C Water Label, Keele, UK.

ONS (2012) Census 2011 result shows increase in population of the South East. Office for National Statistics, Press Release, 16th July 2012. ONS, London.

Ornaghi C and Tonin M (2015) The Effect of Metering on Water Policy Consumption – Policy Note. Economics Department, University of Southampton, Southampton, UK.

Pace R (2014) Managing Non-Revenue Water in Malta: Going Towards Integrated Solutions SMi, Smart Water Systems Conference, London, 28–29th April 2014.

Percili A and Jenkins J O (2015) Smart Meters and Domestic Water Usage: A Review of Current Knowledge. Foundation for Water Research, Marlow, Bucks.

Pint E M (1999) Household Responses to Increased Water Rates During the California Drought. Land Economics, 75(2): 246–266.

Priestly S (2015) Water meters: the rights of customers and water companies. House of Commons Library, Briefing Paper CBP 7342, House of Commons, London.

Rentier G (2014) How smart water meters can help consumers save energy. SMi, Smart Water Systems Conference, London, 28–29th April 2014.

Scottish Water (2016) Scottish Water, 2015–16 Annual Report and Accounts. Scottish Water, Dunfermline, UK.

Sierra Wireless (2014) Unlock the Potential of Smart Water Metering with Cellular Communications.

Slater A (2014) Smart Water Systems – Using the Network. SMi, Smart Water Systems Conference, London, 28–29th April 2014.

Smith A L and Rogers D V (1990) The Isle of Wight Water Metering Trial. Water and Environment Journal 4 (5): 403–407.

Smith H (2015) Universal Metering in Jersey. SMi Smart Water Systems Conference, London, April 29–30th 2015.

Smith T (2015a) How water efficiency can empower people to reduce their bills. Presentation at Water Efficiency, Past, Present, Future, Waterwise Conference, London, 19th March 2015.

Smith T (2015b) Understanding the Customer's Perspective. SMi Smart Water Systems Conference, London, April 29–30th 2015.

Snowden H (2013) Impact of metering non customer supply pipes in Jersey. SBWWI Annual Meeting and Leakage Conference, Intelligent Networks, December 2013.

Symmonds G (2015) The Challenge of Transitioning from AMR to AMI. Presentation given at the SWAN Forum 2015 Smart Water: The time is now! London, 29–30th April 2015.

Symmonds G (2016) Broadening the base: Expanding the potential of smart water. Presentation at Accelerating SMART Water, SWAN Forum, London, 5th–6th April 2016.

Thames Water (2014) Final Water Resources Management Plan, 2015–40. Thames Water, Reading, UK.

Tucker A (2014) Smarts and Water Efficiency. SMI Smart Water Systems Conference, London, April 28–29th 2014.

Tucker A (2015) Smarts and Water Efficiency. SMi Smart Water Systems Conference, London, April 29–30th 2015.

UKWIR (2003) A Framework Methodology for Estimating the Impact of Household Metering on Consumption – Main Report (03/WR/01/4) UKWIR, London, UK.

US EIA (2013) 2009 Residential Energy Consumption Survey (RECS), US Energy Information Administration, Washington DC, USA.

Waldron T, Wiskar D, Britton T, Cole G (2009) Managing Water Loss and Consumer Water Use with Pressure Management. Water Loss 2009 Conference, Cape Town, South Africa.

Water UK (2014) Industry facts and figures 2014. Water UK, London.

Water UK (2015) Industry facts and figures 2015. Water UK, London.

Wallace C (2015) Building Trust with Customers. SMi Smart Water Systems Conference, London, April 29–30th 2015.

Wang U (2015) Water Meters Begin to get Smarter. Wall Street Journal, 5th May 2015.

Wessex Water (2011) Towards sustainable water charging. Wessex Water, Bristol, UK.

Wessex Water (2012) Towards sustainable water charging – conclusions from Wessex Water's trial of alternative charging structures and smart metering. Wessex Water, Bristol, UK.

Westin (2015) City of Santa Barbara, AMI Business Case, Westin Engineering Inc., USA.

Wheeldon M (2015) The beginning of Smart Wastewater Systems? SMi Smart Water Systems Conference, London, April 29–30th 2015.

WICS (2013) Water and sewerage services in Scotland: An overview of the competitive market. WICS, Sterling, UK.

WSA (1993) Water Metering Trials, Final Report. Water Services Association, London, UK.

WWi (2016) Anglian Water targets zero bursts in new trial. WWi, June–July 2016, p.9.

Zpryme (2014) Smart water survey report 2014, Badger Meter, Milwaukee, USA.

Zpryme (2015) Smart water survey report 2015, Neptune Technology Group, Tallassee, USA.

Zur T (2015) Migrating from AMR to AMI. Presentation at the SMi Smart Water Systems Conference, London, April 29–30th 2015.

5

Optimising how we Manage Water and Wastewater

Introduction

The most efficient water and wastewater utilities are those that use the fewest assets to deliver the highest quality service at the lowest cost through optimising their activities. This involves the effective use of data generated by customers' smart water meters (and possibly in the future, smart sewer meters) and blending this with data generated within the utility's abstraction, treatment and distribution systems. It is also concerned with balancing their water and wastewater treatment assets with the current and forecast treatment and handling needs and ensuring they operate in the most effective manner.

In Chapter 3, some examples of optimised water management were presented. Case Study 3.1 highlighted cost savings attained at Northumbrian Water's regional control centre while Case Study 3.3 and Case Study 3.4 considered reductions in non-revenue water reduction at Aguas de Cascais and service improvements through smart meter for Aguas de Portugal. This chapter looks at optimising water and wastewater management in the broader sense.

A utility ought to aim to reduce the asset intensity of its operations while enabling it to deliver the highest service quality at an affordable and financially sustainable price. This combines assuring the security of supplies with building customer confidence in these operations, both in the reliable delivery of drinking water that meets customer and regulatory expectations and the safe removal and treatment of sewage.

5.1 Traditional Techniques and Expectations

Water management has traditionally been risk-averse because of public health concerns and less public tolerance of service failures (or even perceived shortcomings) than for other utility services. Likewise, data gathering has been characterised as being slow, partial, labour intensive, and reactive to events. With most assets being located underground, utilities typically have a limited understanding of their condition or performance. This has resulted in a poor capacity to respond to new challenges through past experiences.

Until a utility's management is able to appreciate how its assets are performing, it is difficult for management to make properly informed decisions about any aspects of their operations that need to be addressed, let alone how to respond to these and to

Smart Water Technologies and Techniques: Data Capture and Analysis for Sustainable Water Management, First Edition. David A. Lloyd Owen.
© 2018 John Wiley & Sons Ltd. Published 2018 by John Wiley & Sons Ltd.

prioritise them. A better informed utility is also able to make a stronger case for investing more to improve and extend its activities. Leakage is a particular concern as it is often perceived as the public face of a utility's performance.

In England and Wales, water leakage was not seen as a priority before privatisation in 1989. Accurate reporting of leakage only emerged in the early 1990s and became a political issue in the wake of the 1995 drought, when Ofwat, the sector regulator imposed targets in 1997 to reduce leakage to an 'economic level' (further reductions would raise overall costs) by 2002–03. Because previous leakage assessments were inaccurate and underestimated actual leakage, leakage rates were perceived to be rising before 1995 (Stephens, 2003). Leakage continues to be problematic for many utilities in England and Wales, but progress has been made; identified leakage rose from 4,781 megalitres (Ml)/day in 1992–93 to 5,112 Ml/day in 1994–95. It was reduced to 3,306 Ml/day in 1999–00 and was at 3,087 Ml/day in 2015–16. Ofwat is proposing that leakage will be reduced by 15% between 2020 and 2025 (Ofwat, 2017).

Leakage or distribution loss ought to be seen as the money being spent on treating and delivering potable water that is not billed for. This means that either more assets and operating spending is needed to deliver the water required, or that water which could be bought by a customer is being lost before it reaches them (Slater, 2014).

A survey of utilities, serving 513 million people mainly in developing economies in 2010 (IB-NET data, Danilenko et al., 2014) found median non-revenue water (NRW) was 28%, compared with 31% in 2000. 29.59 billion m^3 of water was consumed with total revenues of $23.96 billion. If NRW was reduced to 20%, this would provide a further 2.37 billion m^3 of water which in turn could generate an additional $1.92 billion in revenues. Participating utilities in the IB-NET scheme are amongst the better developed in low to medium income countries, the scope for savings through lowering NRW worldwide will be appreciably larger than this.

A survey of 10 NRW, UFW (Unaccounted for water) and leakage reduction programmes in two developed and eight developing economies between 1995 and 2007 gives an idea of the scope for improving the delivery of water. NRW was reduced from 35–61% to 15–37% (n = 5), UFW from 45–52% to 24–43% (n = 3) and leakage from 28–35% to 10–23% (n = 2). These programmes ran from year one to 11 (Ardakanian and Martin-Bordes, 2009). Leakage prevention also has an important role to play in avoiding the need to develop surplus assets (see Chapter 2.3).

5.2 Living in a Real-time World

Information gathering in real-time enables utility managers to reach to events soon after their outset rather than when the impact of the event has eventually been noticed and communicated to them. This minimises the impact of the event both in terms of service disruption and damage to infrastructure. It also enables managers to better understand these events and to improve their ability to predict and respond to such events in the future.

Water asset monitoring involves being able to appreciate how the entire water and sewerage network and their allied treatment facilities are performing at any one time. The more detailed and timely this data the greater is its value. This also involves being able to blend in any applicable external data such as weather, river flow and quality.

When a great deal of data is continually being generated from such services, effective presentation in relation to its importance and highlighting where any interventions is needed. This data is also part of a continual feedback loop aimed at further improving the efficiency of the utility's operations.

5.2.1 Why we Need More Testing – Intensity of Water Use

Water scarcity is often allied with water quality issues, for example saline encroachment of groundwater and the need to maintain the integrity of renewable resources such as rivers and lakes. This requires more detailed data on inland and groundwater quality, both in terms of their availability and quality. Where treated wastewater is reintroduced into river systems or groundwater (aquifer recharge) for indirect potable reuse, this also needs to be monitored appropriately.

The greater the resource intensity, the greater the need for water quality testing; in South East England, 41% of renewable resources are currently being abstracted (EEA, 2009). This has significant implications for inland water quality and means that groundwater levels need to be regularly monitored.

5.2.2 Why we Need Faster Testing – Predict Rather than Respond

The greater the lag between an incident occurring and its being addressed, the more damage can be done. A water leak can damage roads and pavements above it by leaching away the soil lying beneath the hard surface. Sewer leaks can also contaminate water supplies through egress into groundwater and poorly maintained water pipes.

5.2.3 The Role of Domestic Smart Metering in Informing the Utility

The household smart meter completes the picture and is the final element in a smart infrastructure for monitoring water flow through each DMA. It notifies the utility how much water leaves each DMA and is either beneficially consumed or lost through leakage within the customer network. As discussed in Chapter 4, it is also a point of entry for the utility to engage with customers about their water usage.

5.3 Network Monitoring and Efficiency

Water flow monitoring ought to take place from the point it is introduced into the distribution network to the customer, in order to anticipate and react to any deteriorations in asset condition. The same applies to the sewerage network. In the case of sewerage, monitoring also allows a utility to appreciate where rainwater and foul (waste) water actually flow through storm and foul sewers and how combined sewers are in fact performing.

5.3.1 Leakage Detection and Location

Digging up roads to look for leaks in urban areas is increasingly unacceptable, due to the disruption caused. Minimising service interruptions and visible water leakage are tools for improving the customers' perception of a utility's performance and demonstrating that their tariffs are being well spent. Remote and accurate leakage detection enables leaks to be swiftly and accurately located, minimising the amount of digging and disruption needed, or effectively avoided through trenchless approaches.

Undetected leaks are also a long-term impediment to utility performance. Detecting chronic, underlying leakage traditionally requires manual inspections, based upon acoustic detection. The labour-intensive nature of such work means that smaller leaks are usually not detected until a pre-planned, periodic inspection takes place, which may not be for months or even years. This can be addressed through remote acoustic sensing. Three examples of such approaches are outlined below, covering Selangor, Lyon and Milan.

In Malaysia, SYABAS (syabas.com.my) serves 7.5 million people in the state of Selangor. One of its pressing challenges is mains leakage outside the district metering areas. The Echologics (echologics.com) Echoshore acoustic leak monitoring system is being rolled out as a series of nodes which send data to a secure server via local mobile communications networks. Noise loggers by Guttermann Zonescan (en.guttermann -water.com) are placed along the network and acoustic signals are correlated between the adjacent loggers. During 190 days, 1,461 km of mains pipeline were inspected, detecting 154 leaks, 135 of which were immediately repaired. The average leak identified was for 125 m^3 per day (Bracken and Benner, 2016).

Veolia is deploying 5,500 loggers in a trial in Lyon. During the installation of the first 4,400 loggers, more than 260 leaks were newly identified. Loggers are typically placed 30–40 meters apart and the cross correlation of leak identification allows for a higher degree of accuracy in locating the leak. Google Street View is used to visualise there the leak has been located. The automated correlation of readings is significantly more sensitive than unsupported noise level measurement and improved the suppression of false alarms. Data feedback will in time increase the sensitivity of leak location and eliminate more false alarms (Traub, 2016).

The ICe Water project (2013–15) developed a real-time decision-making system for leakage, consumption and flow pressure. This phase concentrated on developing and implementing real-time monitoring, alerts and operational support. It was developed at MM Spa, the Milan Water Utility (medtropolitanmilan.it). MM's Abbiategrasso pilot study optimised pump scheduling, reducing the average pump's energy intensity from 0.443 kw/h per m^3 to 0.388 kw/h per m^3, saving €104,681 in energy costs per annum and the leakage reduction programme saw a 22% reduction in night flow leakage saving a further €111,493. The project had a payback time of 1.5 years (Lanfranchi, 2016).

In each case, the emphasis has been on enabling the utility to address underlying losses in a less labour-intensive manner than was previously feasible.

5.3.2 Assessing Asset Condition

Effective and economical pipe repairs depend on an accurate diagnosis of pipe condition. Water and wastewater assets are mainly located underground which means these assets are poorly understood. As a result, they may either be allowed to deteriorate to an unacceptable degree, or they are replaced before they in fact ought to be, or they could have their operational lives extended.

Water pipe linings are affected by a wide variety of physical, chemical and biological reactions over time, including corrosion, the build-up of protozoans, along with pipe surface erosion and sloughing. Biofilms present in pipes need to be appropriately

Table 5.1 Cost savings with PODDS.

Utility	Mains length	Planned work	PODDS alternative
Wessex	4 km	Swabbing – £490,000	Flushing – £227,000
Wessex	7 km	Swabbing – £530,000	Conditioning – £150,000
Wessex	6 km	Replacement – £2,000,000	Conditioning – £40,000
Wessex	10 km	Flushing – £1,300,000	Conditioning – £40,000
NWG	4 km	Jetting – £300,000	Conditioning – £5,000

Adapted from Boxhall (2016).

managed. For the customer, network deterioration is typically manifested in water discolouration. Utilities need to understand and manage the cohesive layers within pipes and to develop their actual understanding about their pipes, such as their condition, age, material and diameter.

Pipe cleaning is expensive and can only be justified if there will be no significant biofilm regrowth within at least a year. Pipe conditioning can be a viable low-cost alternative. To see where this is the case, PODDS (Prediction of Discolouration in Distribution Systems) predictive modelling can be used (Boxall, 2016) (Table 5.1).

NWG is rolling PODDS out to 923 km of major mains by 2020. An automated pipeline management system covering 350 km of major mains at a cost of £6 million will be in place by 2018. It is anticipated that this will reduce the cost of pipe cleaning from £170 per meter to £17 per meter (Baker, 2016).

PODDS is based on the real or near real-time analysis of water quality, including flow and turbidity monitoring, using long-term time series data. Models can simulate the mobilisation and accumulation of material layers onto a pipe surface, with the regeneration rate being a function of water quality. This means that accurate long-term simulation of biofilm regrowth is possible. When this is carried out within a smart network, it also offers the possibility to manage flows in relation to current and predicted pipe condition.

5.3.3 Water Pressure Management and Leakage Detection

The greater the pressure within a water network, the more water will be forced out of the pipes through any leaks in the system. The effective life of the pipes will also be lowered due to the stress induced upon them. Physical interventions when testing for leaks such as by flexing or banging a pipe will also tend to make the age faster (Dunning, 2015). A less invasive approach is to use a pressure spike, which can be monitored as it moves through the network. Severe pressure transients can damage pipes as well as allowing water from outside the pipe to enter into an area of exceptionally low pressure within the network (Jung et al., 2007). Such transient activity can be caused by a malfunctioning valve and by early identification and replacement, network pressure can be smoothed in order to maximise the operating lifetime of the pipe (Dunning, 2015).

Pressure management can reduce leakage by ensuring that pressure within a distribution system does not exceed its optimal level, especially where demand can change appreciably both during the day and between seasons. Trials with i2Os (i2owater.com) oNet with Severn Trent, Portsmouth Water and United Utilities between 2008 and 2010 resulted in leakage reductions of 26%, 29% and 36% respectively. When 200 systems were installed in Selangor, they reduced leakage by 35,000 m^3 per day and lowered the mains burst rate by 48% (see section 5.3.1). A current contract with Anglian Water, which is seeking to reduce overall leakage by 20,000 m^3 per day from 192,000 m^3 per day in 2014–15 by 2020, has resulted in 40% fewer burst mains and a 35% reduction in leakage. In Guangdong, China, a relatively new water network in the Changping industrial zone managed by GuangDong Water had an 18% reduction in leakage.

There are two broad approaches to leakage detection and monitoring: those which divide a distribution system into district management areas (DMA) and those which look at the whole or part of a network as they feel is best. Case Studies 3.4 and 3.5 looked at the DMA approach in Aguas de Cascais and Aguas de Portugal respectively. The DMA is mainly seen in Europe, especially in France, England and Wales and Portugal, along with Israel, Singapore, Australia, Chile and Brazil, while countries where DMA is rarely used include the USA, Germany and in many developing economies.

The DMA approach has been adopted in recent years for example in India, China and the Philippines. For example, a NRW reduction project serving Manila's west zone (Maynilad Water, mayniladwater.com.ph) managed by Miya (miya-water.com) from 2008 to 2014 combining pressure management and leakage detection in 1,500 DMAs saw NRW fall from 1,580 Ml/day to 650 Ml/day (2,850 to 800 l/connection per day) with 277,000 leaks repaired with the ILI falling from 350 to 40. The improved water availability allowed Maynilad Water to increase its customer base from 700,000 connections to 1,160,000 with improved revenues, average pressure rising from 4 to 19 m and supply time at 24 hours per day instead of 15 (Merks et al., 2017). In 2008–14 Maynilad invested $410 million in NRW reduction including $18 million for Miya resulted in $441 million in additional revenues (Miya, 2015).

The DMA approach generates high quality data, but it is asset intensive. It is also associated with pressure management. Approaches that do not use DMA concentrate on actual leak detection rather than system losses, using inspections and acoustic surveys. Hybrid approaches are also emerging where virtual DMAs (VDMAs) are developed through sensors, meter data analysis and dedicated software systems (Hays, 2017). Case Study 5.3 looks at DMA development in Jerusalem. While the DMA approach does not necessarily need domestic metering, it is appreciably swifter and more accurate when linked to smart domestic metering.

Non-DMA approaches have typically used acoustic leak detection to identify where a leak is occurring along with network inspection for deciding about more general pipeline maintenance and replacement. Acoustic leak detection started with listening for the sound of a leaking pipe and leaks, and has evolved into a number of approaches through data generated by analysing the sounds within the water network and sound pulses. These approaches are increasingly being driven by the need to respond more rapidly to leaks, to locate them more accurately and the nature of the leak. A new approach has been developed by Utilis (utiliscorp.com) using satellite imagery to detect the presence of treated water in the environment by pinpointing its spectral signature through algorithmic analysis and overlaying the data onto maps for utilities to locate where leaks are taking place. Data is updated every 3, 6 or 12 months. When this data was made available to the on-ground leakage teams, their productivity increased from detecting less than 1.76 leaks per person per day to more than 6.1. The intensity of leak detection also improved with 1 leak being found every 0.19 miles per person per day against 1 per 1.9 miles before. The system appears to be more sensitive to larger leaks. A field trial in June 2016 for Grupo Hera (grupohera.it) in Ferrara (Italy) found all large leaks (six, more than 30 litres per minute), 71% of medium leaks (10 out of 14, 5–30 litres per minute), 44% of small leaks (7 out of 16, 0.1–1.0 litre per minute) and 32% of micro leaks (7 out of 22, less than 0.1 litre per minute). Mapping work carried out for Hera across Bologna in 2015 identified savings of 1.5 million m^3 per annum.

5.3.4 Optimising Pumping

Another aspect of pressure management is to ensure that pumps are being used in the most effective manner. Smart systems for optimising pumps have a significant potential to reduce energy consumption in the water sector amongst other applications. According to Grundfos (Riis, 2015), pumps account for 10% of global energy consumption and 90% of pumps currently in use have not been optimised. Pump operation optimisation includes using sensors to transmit pressure data within the distribution network to support an intelligent pump system using algorithms; these establish the ideal pressure required within the distribution network at any one time in real-time.

Since energy accounts for most of a pump's overall costs, it makes sense to consider pump efficiency, deployment and management as a chief priority when developing new distribution systems or managing extant assets. Energy use needs to be factored into pump management, as in terms of a pump's life cycle costs, 5% is spent on buying the unit, 10% on servicing and maintaining the unit and 85% on its power consumption. Examples cited in this chapter (Case Study 5.5) show how energy can account for 20% (Fargas-Marques, 2015) to 24% (Carvalho, 2015) of water utility operating spending. In a typical water distribution system, pumps account for 89% of energy consumption, along with 6% for offices, systems and lighting and 5% for filter backwashing (Bunn, 2015).

Derceto's (now part of Suez) Aquadapt software is designed to lower a water utility's energy needs by optimising the timing of its pumping activities in relation to resource availability and electricity tariffs. In 2000–01, energy costs for a trial involving 60% of Greater Wellington Water's network in New Zealand lowered energy costs by 12%. The service was rolled out to the rest of the network in 2008. WaterOne (weaterone.org), serving 400,000 people in Kansas City, USA has lowered its energy bills by over $1 million per annum through the management of 32 pumps and six flow control valves, reducing peak electrical demand by up to 4 MW.

Pump efficiency analysis is based on measuring the suction and discharge pressure, along with the pump station's total flow and energy consumption. This data ought to be verified as until recently, it was of a widely varying quality. Pump efficiency analysis also needs to factor in when two or more pumps are working together. This provides a quite different energy profile to that of a single pump. The next step for pump efficiency will be benchmarking pumps between individual pumps and groups of pumps, between pumps at the water utility level and finally, at the international level.

Running pumps in the most efficient manner requires variable frequency drives for pump speed control and a multi-pump controller for overall pump management within a network. When pump system management is linked with active water network pressure management, this can result in up to 20% less energy being needed along with reducing network distribution losses by up to 20%.

Optimal pump usage levels can result in 11.5% improvements in energy efficiency (Bunn, 2015). This is related to the amount of pumping needed and the deployment of a group of pumps so that whichever pumps are being used at any one time are used in an effective manner. For example, by switching from one small pump to two small pumps, or from three small pumps to two large pumps, when this results in pumps being used at their most efficient level.

5.3.5 Dealing with the Data

One of the challenges for utilities is the ability of their current or planned data systems to manage the increased volume and velocity of data. Firstly, specific vendor-developed MDM (meter data management) platforms are not usually designed to typically handle multiple data sources. They are also typically AMR or AMI specific, with no capability of cross-platform interoperability. Also, there are few utility systems that are designed to actually use the highly specific data that is produced by these platforms. Systems that are designed to operate with two data reads per month are unlikely to have the ability to handle the 720 (once every hour) to 2,880 (once every 15 minutes) customer reads that are capable of being generated by AMI systems (Symmonds, 2015).

Therefore the amount of data being generated is of a different order of magnitude than before and this needs to be managed if it is to be of use. For example, the city of Cachoeiro de Itapemerim in Brazil, is served by a utility with 55,309 metered connections and a water loss of 185.6 litres per connection per day (the average for Brazil is 366.9 litres per connection per day). Within the system, there are 1,956 data points divided between the 20 DMAs. Each data point generates 5.18×10^6 units of data a day or 1.01×10^{10} units of data a day overall, some 3.70×10^{12} units of data each year (Sodeck, 2016). This data needs to be cleaned, to remove gaps, zeros, peaks and constant values through the effective use of previously collected data. In this case, deviations from expected system performance over the previous 30 days are reported every 30 minutes and presented in graphics, along with other potentially relevant data (for example, pressure, reservoir levels) and minimum night flows.

5.4 Drinking Water – Quality

A water system's integrity matters to its users, as dramatically demonstrated by the lead contamination problems encountered in Flint, MI, USA since 2014 (Dingle, 2016). It is evident that one of the chief drivers behind the subsequent crisis was the poor quality of water testing across the town's facilities and networks.

Drinking water quality is driven by the World Health Organization's (WHO) 'Guidelines for Drinking Water Quality' the 4th edition being published in 2011 and the 5th edition due to be published in 2020. Effective monitoring of water quality through the network identifies potential areas where the distribution mains are deteriorating and where this impacts water quality. Real-time monitoring also enables the utility to treat its water to ensure that all applicable quality criteria are satisfied without the need for excessive treatment. Minimising chlorination for example, improves the taste of the water, while saving on chemical costs.

5.4.1 Drinking Water – Potability, Aesthetics and Public Confidence

Rising expectations about service delivery, as well as aesthetic and public health standards impose their own costs, both on monitoring and in satisfying customer demand. Public willingness to pay for services is linked to service quality, especially in the perception of service reliability.

Where there is little public trust in service delivery, alternative approaches are increasingly being adopted. This is discussed in some detail in Chapter 6. These include point of use (PoU) and point of entry (PoE) household water treatment units and bottled water, irrespective of the latter's actual quality. This is usually most evident in developing economies, although in Mexico and increasingly in parts of the USA, such as California, spending on bottled water spending can exceed water utility revenues. Here, consumers are spending on water, but the money is not going on infrastructure or service development.

5.4.2 Going Back to the Source – Catchment Management

Catchment-based water management is based upon a full appreciation about the characteristics of the water flowing through each water system from its origin to its consumption and discharge. By working with farmers, managers of upland assets and other stakeholders, downstream water quality can be improved and flood resilience boosted through a greater capability to absorb exceptional rainfall upstream (Indepen, 2014) with the potential to improve agricultural efficiency, lower flooding costs and decrease costs related to drinking water, wastewater and the quality of inland waterways.

The sooner distribution problems can be detected, the less their impact and the more precisely their location can be detected. For water utilities this is in part being driven by external factors, as climate change causes inland water temperatures to rise and greater variations in seasonal flow create conditions better suited for the growth of waterborne infections such as cryptosporidium, which can enter the distribution system from upland water sources.

Upstream catchment management to date has concentrated on the potential for physical interventions to improve downstream outcomes. Monitoring has been reactive, with an emphasis on long-term outcomes. As hardware costs fall, migrating monitoring upstream allows a greater focus on those areas where upstream data can provide an early warning about potential issues that may arise downstream, enabling operators to assess and address them at the earliest instance. Smart catchment management includes considering how to alleviate the rising conflicts between agriculture and utilities for water resources where irrigation agriculture is practised.

5.5 Water Utilities and the Wider Environment

A sewerage utility's obligations extend to the discharge of post-treatment effluent into inland or coastal waters. More sensitive and rapid detection of pollution incidents and effluent loading mean that the utility has to be in turn more responsive to any current or potential effluent discharge issues. More rapid detection capabilities enable utilities to respond more effectively, ideally before the incident has any significant consequences.

5.5.1 River and Ground Water Quality Assessment

Water treatment can be optimised when the status of the raw water at the inlet point is properly monitored and appreciated. This extends to how the water may be affected by external factors, including working with the appropriate environment agency or inland water management entity to appreciate the quality of the water as it flows through the network. This includes river water flow, temperature, turbidity, BOD and the presence of various potential contaminants. The earlier any perturbations from accepted norms can be detected, the greater the chance of minimising the effect of pollution incidents.

5.5.2 Flood Detection and Management

While utilities typically have a limited role with regards to preventing and managing flooding incidents, they are very much associated with them, being in the public eye as the day-to-day providers of water and sewerage services in any affected area. The frequency and intensity of flood events is increasing in many areas and this is in part being exacerbated by urbanisation, as in urban areas, 85% of rainfall becomes surface run-off, which has to be absorbed by drainage systems. The role of external factors such as soil moisture content prior to heavy rainfall and surface run-off have not tended to be adequately factored into flood risk monitoring until comparatively recently.

Flood costs in the European Union averaged €4.9 billion a year from 2000 to 2012. This could increase to €23.5 billion by 2050 with major flood events increasing in frequency from an average of once every 16 years to once every 10 years by 2050 (Jongman et al., 2014). Globally, without adaptation, assuming a 25–123 cm global sea level rise by 2100, 0.2–4.6% of the global population is expected to be flooded annually resulting in annual losses of 0.3–9.6% of global GDP. The global coastal protection costs will be in the range of $12–71 billion per annum (Hinkel et al., 2014). The economic damage from floods in 1995–15 was estimated at $662 billion, at $216 million per flood (Jha et al., 2011) while economic losses by decade have been estimated to have increased from $5 billion in the 1950s to $40 billion in the 1980s and $185 billion in the 2000s (CRED/UNISDR, 2016). There are two 20-year figures for recent global economic loss by flooding, with a range of $10.8 billion (CRED/UNISDR, 2016) to $19.8 billion (Jha et al., 2011) per annum.

Flood management is usually concerned with large-scale 'hard' defences, rather than considering flood avoidance and amelioration and the more effective deployment of defences where they are most needed. In England for example, £1,500 million per year is spent on land management that is neutral or worse with regards to flooding vulnerability against £418 million on measures designed to ease flooding events, £269 million on hard flood defences and £613 million on post-flood repairs (Wheeler et al., 2016).

Flooding incidents such those as seen in England in January 2016, when newly installed flood defences performed to expectations but were overwhelmed by exceptional river flows, demonstrate how climate change is challenging our understanding about how the nature of extreme weather events and how this affects the way flooding occurs and can be managed.

5.5.2.1 Smart Flood Management

Two types of inland flooding take place; pluvial flooding (rising river and/or water table levels) and flash flooding (intense discharge of water into rivers and streams). Coastal flooding also involves seawater, sometimes allied with storm events. Smart flood management can be divided into seven stages. Firstly, mapping and analysing the vulnerability to flooding in a given area, such as a river basin. Responses can then be developed for these vulnerabilities, through flood resilience measures, to an appropriate degree. This can involve flood prevention (increasing the natural or engineered absorptive capacity within or before a flood zone) as well as developing flood defences on both a property and area level. Water flows and the weather are then monitored in real-time and the data is fed back to the maps to improve their accuracy and extend their predictive ability. The maps in turn are continually updated for changes that may affect the flow of water through the area, along with changes in water levels and climate patterns, to highlight current or emerging vulnerabilities. From here, potential flooding events are forecasted, in order to trigger the deployment of flood defences and warn people in the affected area as far in advance as possible. In some urban areas, people can also be warned to alter their water usage (baths, washing machines and so on) during critical periods to lower the impact of storm sewer flooding. Sewer flooding is examined in section 5.7.1.

Smart flood management covers the development of a real-time and localised appreciation of current and future flood vulnerabilities and the most effective responses to these at the local and regional level. Case Studies 5.9 (Portsmouth, UK) and 5.11 (Bordeaux, France) examine how two smart flood warning and management systems have been established at the catchment level. The following two examples outline the potential to identify such vulnerabilities at the local and indeed, household level.

PyTerra (pyterra.co.uk) is an early stage company (owned by Concepture Limited) seeking to develop smart water enabled risk management systems for the UK market. Water-risk mapping is traditionally based on historic data and concentrates on single risk issues. PyTerra aims to integrate a broad spectrum of water risk data onto a common platform based on GIS, satellite imagery, Big Data and hydrological modelling through its in-house optimisation system. The output is designed to be updated as and when any new and applicable data becomes available.

Access to comprehensive and comprehensible data about water-related risk is a constraint both for risk management and for considering the viability of planned developments. This is particularly the case for stakeholder engagement where appreciating the nature of risks can be difficult to communicate. The PyTerra system presents a single, unified view of events through a base layer of open data which will be freely available to all stakeholders so as to seek to circumvent this. Premium data tools are then available by subscription to parties with a commercial interest. Their smart water mapping allows users to interact with the specific data they are looking for. Applications envisaged include irrigation for farmers, flood compensation work for housing developers and Water Framework Directive (2000/61/EC) compliance tools for environmental regulators and agencies.

IMGeospatial (imgeospatial.com) is owned by Intelligent Modelling, developing smart products for the flood management market. A digital terrain model (DTM) uses currently available data (including maps, satellite imagery and planning applications) to create a Pluvial Hazard Map that incorporates features such as kerbs, hedges, fences, walls and bridges, along with buildings and roads, which modify surface water flows in order to create a model of the area's flood risks. It is designed to be self-updating through remote access to new sources of relevant data (Evolving DTM). When changes such as house or road building that can affect flood vulnerability are recorded, alerts are created. This can be applied during the planning process to consider the potential impact of proposed developments. The DTM is then integrated within urban water flow paths such as ponds, sinks, connecting paths and sewers. From this, flow pathways in flood-prone areas can be identified and the relationship between surface and sewer flows appreciated, to predict where actual flooding events may occur in relationship to the network's capacity and different weather events. By simulating water flows through modelled systems, clients are able to understand how the actual system will perform under a range of conditions. As further data becomes available relating to real and simulated events, and is integrated into the model, IMGeospatial aims to develop this into an active flood warning system.

Another approach is to consider the potential for sustainable drainage systems. These both include the sustainable drainage systems (SuDS) approach in urban areas and its broader application in rural areas, combined with catchment management. SuDS approaches are typically applied on a localised scale as a complement to other urban drainage systems. The difficulty facing the broader deployment of SuDS lies in the variety of approaches available and the amount of data generated when comparing them. The SuDS Studio developed by Atkins (atkinsglobal.co.uk) was used to examine various approaches for 1,900 km^2 of land within Anglian Water's catchment area. The system uses GIS data to map potential SuDS areas while ensuring potential conflict zones such as listed buildings, flood zones and incorrect topography are avoided. Within the area analysed, 13.5 million potential projects were identified, 5.5 million of which were cost-effective (Todorovic and Breton, 2017). From here, more detailed analysis can be carried out, based on a database with the degree of detail that was not feasible with non-smart approaches.

Storm surge monitoring systems are also being developed for coastal areas. 'Stormy' a storm surge forecasting system developed by the National University of Singapore combines weather forecasts with wind and sea surface pressure data and sea levels to generate a daily graphically based six to seven day alert of storm surges from 0.3 to more than 0.5 meters. The system covers the Straight of Singapore in the South China Sea (Luu et al., 2016), and successfully forecast two storm surges in 2013 and 2014 with an accuracy of 0.05 meters. The system is currently being upgraded to incorporate tidal oscillations and to extend its geographical coverage.

5.5.3 Bathing Water Monitoring

Monitoring of bathing water quality was traditionally reactive, with a lag of some days or weeks before test results were displayed at the resort in question, an overall assessment being made about each beach's quality at the end of the year. With the EU's revised bathing water directive (2006/7/EU), data is made easily available on dedicated websites with a minimal delay from testing. This allows for a continual monitoring of bathing water quality and the ability to highlight any problems as they occur, so that potential visitors can react to events and for authorities to close a beach if necessary. As bathing water quality is affected by rainfall (rainwater flushing sewage from combined sewer overflows) timely monitoring and data dissemination is important.

While the directive was adopted in 2006, some time before the potential for smart water management to include areas such as this, the need for real and near real-time data collection and dissemination, especially at remote locations such as CSOs have meant that as the directive became fully effective in 2015, the evolution of its requirements and the capabilities offered by smart water have aligned to their potentially mutual benefit.

5.6 Wastewater and Sewerage

5.6.1 Sludge Condition and Treatment

Sewage treatment works most effectively when it is carried out within optimal operating parameters based on knowing about the wastewater which is flowing towards the treatment works, both in terms of volume and its effluent loading.

Reactive responses to changes in operational parameters at a wastewater treatment works carry the risk of compliance failure. For example over-treating for nitrification (during off-peak periods) is a waste of energy and can raise total nitrogen discharge, while under-treatment (during storms) may result in high ammonia discharges. Real-time monitoring measures the plant's actual treatment capacity, along with the incoming load and the effluent quality. Two control loops are driven by the data; a feed forward for reacting to any changes and a feedback to correct these changes before the effluent is discharged. Trials at a 250,000 PE (60,000 m^3 per day) WWTW saw a halving in ammonia discharge at peak times and a 10–20% saving in the energy needed for aeration. The use of detailed monitoring makes the treatment process transparent for the first time, allowing for its effective management. This approach can be deployed in new builds or retrofitted to existing systems (Haeck, 2016).

5.6.2 As a Renewable Resource – Water and Wastewater Reuse

Wastewater is a resource, rather than something that needs to be treated and disposed of. Creating value through water, nutrient and energy recovery from wastewater also enhances the cash flow needed for operating and developing wastewater treatment systems. Water can be returned to the domestic network directly (direct potable, most notably in Windhoek, Namibia) or indirectly (indirect potable, for example in Singapore where NewWater is discharged into reservoirs where it is mixed with river water and subsequently treated) and direct non-potable sale to industrial customers, which is being widely adopted in Australia, India and China, where non-potable water revenues to industrial customers can effectively underpin the project. Nutrient recovery is being driven by limited fertiliser stocks and energy recovery is playing an increasingly important role in reducing utility energy costs and lowering their carbon footprints. It is necessary to tightly control various stages of the waste treatment process to optimise resource recovery, in particular for sludge digestion.

5.6.3 Storm Sewerage Overflow Detection and Response

Ensuring that the foul sewage network and storm sewerage systems effectively interact and that the wastewater treatment works are not overloaded is another priority.

The priority for storm sewerage flow monitoring and modelling is early problem detection and prevention, based upon identifying blockages, polluting CSO overflows and internal and external flooding before they impact a system's performance. This calls for the right forms of autonomous telemetered flow and level monitors in the right places. For pressure depth measurement, ultrasonic level sensors are preferable except where circumstances (for example, small chambers or high surcharge levels) mean that pressure sensors work better. A significant number of monitors installed in the UK during 2005–2015 are already failing because their performance has not been monitored and they have not been maintained. Effective monitoring means using units that do not require maintenance for at least 10–15 years, and which also automatically assess and report their operating condition.

Sewer blockages typically start with the build-up of material in a storm sewer network during a dry spell. Rainfall causes a blockage as the material is shifted through the pipes and water build-up and flow-back causes a pollution event. Further rainfall will clear the blockage and when the weather is dry again, the cycle resumes.

This can be addressed through predictive analysis, based on combining expected events with what is going on at the time. For example, weather report analysis in the control room is used to generate trigger alerts, sites are prioritised, maintenance teams respond and update the system, and this information is blended with new weather data as it becomes available. Yorkshire Water used predictive analysis in 2014 and pollution incidents fell by 21% as a result through incident prevention (Harrison, 2015).

Storm overflows, especially in potentially sensitive waters, need to be monitored and managed, including pathogen strategies for storm waters. Pathogen reduction in water or effluent is defined as the ultraviolet (UV) dose needed to reduce the presence of the pathogen by, for example, 99.9% (a '3 log kill') to a specified concentration (units of the pathogen per 100 ml). UV dosing is validated through a series of field trials taking into account actual flow rates and the number of UV modules to be used (Dinkloh, 2016). Chichester Harbour in the UK is designated under the EU Shellfish Waters Directive (2006/113/EC) and stormwater flows needs to be appropriately disinfected. A Wedeco Duron (xylem.com) stormwater UV system using 10 UV modules down a single outflow channel was deployed by Southern Water in March 2014 to treat a flow of 1,086 m^3 per hour of stormwater. By monitoring the UV transmission and intensity, the operator can indirectly measure the applied UV dose and the effective disinfection performance of the facility. Real-time, continual monitoring of the UV dosing ensures that discharges comply with the Directive.

Broad adoption of smart sewer monitoring remains at an early stage. In February 2017, Severn Trent of the UK awarded a systems-wide monitoring contract to WWM (hwmglobal.com) for 700 Intelligens Flow monitoring systems (sewage velocity and depth), 3,000 Intelligens WW and SonicSens ultrasonic sensors (in-pipe sewage flow) and 1,130 Intelligens Flood Alarm systems (sewer level alerts). The units communicate through a GPRS modem and can be remotely upgraded and recalibrated and are designed to operate autonomously for a minimum of five years. This is believed to be the largest project of this nature to date (Water Active, 2017).

5.6.4 Wastewater as a Public Health Monitoring Tool

Wastewater is a potential source of information about people's health and habits. What goes in goes out; ethinyl estradiol (EE2) in birth-control pills has disrupted male fish endocrine systems in inland waters which receive treated effluent (Owen and

Jobling, 2012), reducing their fertility. Zuccato et al. (2008) found that by measuring the concentration of benzoylecgonine (a compound only found in the urine of cocaine users) they estimated that 40,000 doses of cocaine were being consumed each day in Italy's Po Valley, rather than the previous assumption of 15,000 doses per month. Pharmaceuticals discharged from wastewater treatment works are also of increasing concern (Kostich et al., 2014; Liu and Wong, 2013). Information about the material flows within wastewater and what is carried by these, especially regarding genetic data is now obtainable due to the convergence of genome sequencing, the ability to record DNA data and the ability to analyse, transmit and integrate this information at speed and at a reasonable cost.

One area of particular interest is the early detection of disease. Symptoms of a disease usually become noticeable at some point after the underlying infection has started. The incubation period of these diseases may extend over several days or even weeks. Where DNA or metabolites are discharged before this, early warnings about the incidence of these diseases may become feasible before external symptoms emerge. The ideal would be for biosensors to allow the real-time detection of viral outbreaks. Being able to detect such changes would also enable epidemiologists to better understand how disease develops and spreads across a city. This would in effect become a real-time revisiting of John Snow's 'Ghost maps' that enabled the location of the source of the 1854 cholera outbreak in London to be identified and isolated (Johnson, 2006). Likewise, monitoring enables public health researchers to appreciate the actual use of all types of drugs (Ratti et al., 2014).

'Underworlds' is a smart sewer monitoring system developed by the Massachusetts Institute of Technology's MIT Senseable City Lab (underworlds.mit.edu) with the aim of near real-time data analysis of the composition of a city's sewage. One ambition is to see if each city has a distinct biological signature and to see how this data can be used for developing near real-time public health strategies (Common, 2016). In reality, the cities that need such services most of all, will be those worst equipped to adopt these approaches, due to the limited extent of their sewerage networks. It is those who live in informal settlements who would benefit the most from an approach such as this, but with some exceptions (for example condominial sanitation in Brazil and the Orangi Pilot Project in Karachi, Pakistan), informal settlements can be characterised by the absence of sewerage networks.

The cost of monitoring for the early warning of epidemics would need to be balanced against the costs avoided both in economic and human terms by being able to respond to disease outbreaks in a swifter and more proactive manner. Again, all manner of outcomes may emerge, both expected and unexpected. This could be used to detect the consumption of drugs, both legal and illegal. This would raise a considerable number of questions regarding privacy and civil liberty. These will be considered in Chapter 9.

Another approach being developed is a smart loo which recognises the DNA sequences of individuals in a household and carries out biomarker and microbiota analysis. A simpler version developed by Japan's Toto caries out urine analysis. Size, cost and speed are challenges here, if such approaches are to be deployed to any great extent (Ratti et al., 2014). Currently, considerations mean that these would be devices purchased by early adopters. There may be personal circumstances where early detection of certain conditions would merit the necessary outlay, especially in countries such as the USA where healthcare costs are high.

5.6.5 Smart Sewerage Capacity Optimisation

Sewerage level monitoring and warning systems enable utilities to prevent pollution incidents both from the discharge from the combined sewer overflow (CSO) and in ensuring that there are no overflows (and blockages) within the sewerage network. The smart CSO utilises an overflow monitor, linked to a remote emergency shut-off valve and a control panel and rain gauges. This becomes part of a broader folio of flow management approaches, integrating weather data, along with actual and predicted rainfall, and the performance of the storm sewer network to the drainage basin.

Sewage pumping stations can be optimised through flow monitoring and management. For example, variable frequency drives allow power consumption to be optimised and performance related to blockages and a constantly updated network cleansing cycle. Sensors detect air in the pump chamber and gas formation due to chemical reactions along with diagnosing blockages in the suction area and monitoring system pressure. Pump condition monitors hold system memory related to a real-time clock with an Ethernet interface for data logging.

Within a network, it is necessary to identify critical areas for work prioritisation. Likewise, works need to be integrated with weather data to ensure they can be postponed during weather critical events. Looking forward, the main opportunities lie in the evolution of more sensitive and fast reacting sensors, and relating these to real-time network control and the application of hydraulic models (Kaye, 2015), to current and predicted weather conditions and other events.

Where a community is upgraded from septic tanks to sewerage, retaining the septic tanks' storage capacity offers considerable potential for optimising network and treatment efficiency, especially in terms of treatment needs. Integrating sewerage and septic tanks has been demonstrated in Australia. South East Water served the Mornington Peninsula in Melbourne. It is one of the city's most prosperous suburbs and its houses have traditionally been served by septic tanks. Due to contamination from sewage leakage, a sewerage system was proposed at a cost of A\$ 507 million for the 25,000 properties (A\$ 20,280 per property). The main challenge here is peak usage, due to the seasonal nature of the property occupancy (GWI, 2016a).

A smart sewerage management system was developed through using the extant septic tanks to hold sewage back, so as to smooth flows into the sewerage system. A proprietary One Box control was fitted to each property, which monitors the sewage flow into the septic tank, the sewage level and how and when it is released into the sewerage network. With the sewer pumps running on average for 8 minutes a day, there is room for their flexible use. The One Box system also identifies blockages and where the septic tank has inadvertently been connected to a storm sewer. The system thereby allows for active sewerage management at the household level. When rainfall is anticipated, sewage can be held back in the septic tanks until the rainfall-elevated flows through the stormwater network have eased. The lowering in peak flows meant that the system was installed for A\$ 255 million (A\$ 10,200 per property), or 51% of the anticipated cost, while delivering a more efficient and closely managed service (GWI, 2016a).

Smart sewerage involves the integrated monitoring of sewage flow and other operational parameters across the sewerage and sewage networks and treatment systems, as well as the discharge points such as CSOs, using a systemic understanding of the behaviour and performance of the entire sewage system.

Event duration monitoring (EDM) covers the monitoring and recording of when discharges from CSOs and other permitted storm overflows occur and how long they last for. There are over 15,000 permitted CSOs in England, 89% discharging to inland waters, 10% coast and estuary and 1% to groundwater. Between 1995 and 2015, 6,800 CSOs were improved (Hulme, 2015). Climate change and urban growth are set to increase flood-related sewer volumes by 51% by 2040. Permitted CSOs need to be appropriately managed and the Environment Agency has asked the water and sewerage water companies to monitor the 'vast majority' of their CSOs by 2020 (Hulme, 2014). The CSOs that need monitoring are those where overflows occur more than once a year and at amenity sites or where covered by the EU's Habitats Directive (94/43/ EEC). This is also becoming the case for CSOs discharging into designated bathing waters in the EU due to the standards required to comply with the revised Bathing Waters Directive (2006/7/EU) from 2015. Monitoring with EDM involves volume measurement and telemetry where appropriate. During AMP6 (2015–20) 8,784 CSOs are being monitored. For designated sites (Sites of Community Importance), this means monitoring a spill at two-minute intervals (1,405 CSOs, with volume monitoring where appropriate) and informing all local and national authorities as necessary. For less sensitive areas (7,379 CSOs), a 15-minute monitoring frequency is seen as appropriate (Hulme, 2014).

5.7 Avoiding Surplus Assets

The examples above have mainly considered what smart networks can achieve in terms of improving efficiency. Smart approaches on extant and planned networks and facilities can also defer or reduce the need for new assets.

5.7.1 Making the Extant Networks Deliver More

If extant assets can be effectively deployed to meet a utility's current and future demands, building new assets can either be postponed or not needed at all. It was noted in section 5.1 that a leakage reduction programme is set to prevent the need for substantial new desalination capacity in Riyadh. In 5.6.5 it was shown how extant assets (septic tanks) could be integrated into a new sewerage network to lower the system's peak treatment capacity. Smart approaches are used to identify exactly what needs to be done under each utility's particular circumstances.

The examples below (water in Australia, sewer flooding in the UK and sewerage in Australia) show how demand management, understanding flood vulnerability at the property level and making the fullest use of a sewerage network's capacity can drive down the costs of maintaining, enhancing and expanding service infrastructure.

Demand management cuts water costs and can obviate the development of new assets when extant assets can continue to deliver a suitable level of service. Examples in Australia of demand management (Beal and Flynn, 2014) include the reduction of monthly peak demand by 10%, allowing the deferral of A$100 million in new infrastructure for four years, a Net Present Value saving for the utility of A$20 million. In a second case, deferring a A$20 million water treatment works upgrade by seven years after demand growth was reduced, represented a capital spending savings of A$7.9 million, and deferring a A\$5 million pipeline upgrade for five years realised a capital efficiency saving of A\$1.6 million.

Making a vulnerable property safe from sewer flooding is expensive. The mean average cost of five difference approaches to preventing sewer flooding in the UK ranges from £15,000 to £58,000 per property (Keeting et al., 2015), so sewer flooding prevention and management strategies should concentrate on those houses with the greatest vulnerability. The InfoWorks CS sewerage and flood risk assessment model developed by Innovyze (innovyze.com) subsidiary MWH Soft was combined with 16 flow monitors to examine 916 properties (out of 4,500 in the area) in a part of Blackburn, UK that were considered at risk of sewer flooding. In consequence the At Risk register was reduced to 118, with a higher degree of confidence for flood risk prediction and lower insurance surcharge risk (Innovyze, 2010).

Yarra Valley Water served part of Melbourne, Australia. The suburb of Mernda-Doreen is forecast to grow from 6,000 properties in 2008 to 20,000 by 2030. InfoWorks CS was used to look at the area's future sewerage requirements and identified ways of optimising the extant sewerage system for A\$1 million rather than the A\$10 million sewer as planned (Innovyze, 2009). InfoWorks CS was used to evaluate a real-time control system in Bordeaux, France and Ottawa, Canada to minimise the capital spending needed to meet wastewater management standards. The system enabled savings in the new infrastructure needed of 67% in Ottawa (\$65 million) by avoiding the need for a new sewer tunnel while improving the capture of rain-generated wastewater from 74% to 91%. A cost saving of 63% in the Louis Fargue basin of Bordeaux (€62 million) was attained through improving the network's ability to store wastewater (Innovyze, 2008).

5.7.2 Efficient Deployment of Meters and Monitors

Metering and monitoring are not ends in themselves, they are a means for generating the data necessary for the desired outcomes, and little is gained from surplus monitoring devices or data. Optimal placement of water sensors can be attained through using the Darwin Sampler (bentley.com) programme for pressure loggers, water quality monitors and for flow monitors. This can either be achieved by minimising the number of samplers needed within a district meter area or to maximise the coverage from a set number of sensors. For water quality sensor placement, the Darwin Sampler found a diminishing return from extra sensors. One sensor achieved 30% coverage within a DMA; four gained 50% coverage; 15 reached 75% with 90% coverage from 40 sensors. After this, no additional coverage was attained from extra sensors (Zheng, 2015).

For example, instead of 22 water system loggers being placed by the traditional rule-of-thumb approach, using optimised placement, a coverage rate of 96% within the same area was achieved with six loggers. In the case of hydrant flow testing and flushing rate coverage, eight hydrants selected by experience resulted in a 32% flushing rate coverage, against 66% coverage when the hydrants were selected by a Darwin Sampler (Zheng, 2015).

In addition, the Darwin Calibrator can be used for network model calibration. This includes detecting long-lasting and hard-to-find leaks along with new leaks and to identify a small number of unknown valve settings among thousands of valves within a network.

In a trial 75 extra sensors were deployed by United Utilities (UU, UK) in 17 DMAs to identify the optimal number for network coverage using a Darwin Sampler. The desired network coverage (at least 80%) was reached with five sensors per DMA, with 96% coverage from 10 and 100% from 20. UU is currently looking at the macro-location of leaks in standard sampled DMAs and comparing this with their micro-location in over-sampled DMAs using event recognition systems developed by the University of Exeter. This is to be allied with a leak location system (LLS) using statistical process control for the comparison of pressure-related data over time. Initial results point to improved identification and location of leaks (Romano et al., 2016).

Using a Darwin Sampler to minimise the number of sensors needed within a DMA is an example of a virtual DMA approach (see section 5.3.3). UU started to use DMAs in the 1990 and their aim was to upgrade them with smart capabilities while avoiding excess sampling assets. As United Utilities has approximately 3,000 DMAs, this represents a significant saving for data logging and makes the wider application of more detailed data collection more attractive. Another trial with United Utilities covering four DMAs with 10,274 households and 126 km of pipelines looked at the potential for Darwin Sampling to improve leakage detection. A traditional network sweep-through, using an acoustic device took 353 man-hours. In contract, a locate-pinpoint approach (the model is used to locate leaks and the device to pinpoint them) took 132 man-hours, and brought about 29–44% leakage reductions. In terms of input and results, it was seen as being four times more efficient than the acoustic approach.

Berliner Wasserbetribe (bwe.de) a utility serving 3.7 million people in Berlin has 7,917 km of mains and 150 pumping stations. Network pressure is managed by 50 sensors, in order to minimise the amount of data the utility needs to handle (Freyburg, 2017).

5.8   Case Studies

Eleven case studies are presented below, six on water distribution, three on sewerage and two covering flood warning and management.

5.8.1   Case Study 5.1: Fast Action Leakage Detection in Copenhagen.

5.8.2   Case Study 5.2: Data Logging and Network Optimisation.

5.8.3   Case Study 5.3: Leak Detection and Management in Jerusalem.

5.8.4   Case Study 5.4: 'Mapping the Underground' for Locating Utility Assets.

5.8.5   Case Study 5.5: Energy Efficient Pumping in Spain And Brazil.

5.8.6   Case Study 5.6: Smart Water in Malta – The System.

5.8.7   Case Study 5.7: Wireless Enabled Sewerage Monitoring and Management.

5.8.8   Case Study 5.8: Monitoring for Sewer Overflows.

5.8.9   Case Study 5.9: Flood Warnings and Event Management.

5.8.10   Case Study 5.10: Sewerage Monitoring in a Remote Community.

5.8.11   Case Study 5.11: Flood Warning and Management in Bordeaux.

5.8.1 Case Study 5.1: Fast Action Leakage Detection in Copenhagen

The Danish government has actively sought to encourage efficient water management for some decades. In 1994 a water supply tax was introduced to incentivise water utilities to minimise leakage. Since 1998, where leakage exceeds 10% of the total water supplied utilities are levied DKK 5.0 per m^3 lost above this amount (Lambert, 2001; OECD, 2008), subsequently rising to DKK 6.13 per m^3 (Merks et al., 2017). Although the charge has had a material impact on NRW in Denmark, 12 of the 48 reporting utilities in 2014 still had a NRW of above 10% (Reschefski et al., 2015). In 2002, universal water metering was introduced, followed by double flush lavatories in 2005 and rainwater reuse from 2008. Water and sewerage services in Greater Copenhagen are provided by HOFOR (hofor.dk). In Copenhagen, per capita water consumption has fallen from 171 litres per day in 1987 to 135 in 1995 to 108 in 2010 and 100 in 2015 (Skytte, 2016).

High tariffs (the combined water and wastewater tariff for Copenhagen was $6.21 per m^3 in 2016, the world's highest) mean that the economic level of leakage is appreciably lower than for example the UK or USA. Denmark's NRW has been between 8.1% and 9.6% nationally between 2010 and 2014, with a downward trend since 2011. The Danish utility ILI scores at 0.1 to 1.6 are amongst the lowest seen in the world. Copenhagen is above this with an ILI of 2.5, but this is still classified as an efficient network with an NRW of 6% (Reschefski et al., 2015; Pedersen and Klee, 2013).

Pipeline renewal in Copenhagen is at 0.9% pa for 2005–2014 and 76% of pipelines are over 60 years old (Pedersen and Klee, 2013), demonstrating that older networks can be operated efficiently if appropriately managed. Early leak detection also avoids the costs associated with damaging roads and other infrastructure. The difference in cost avoided can be 50–200 times greater than the typical £2,000 cost of a leak if repaired in good time (Fisher, 2016).

Leak monitoring used to be carried out by moving Permalog+ acoustic detection units through HOFOR's water network across three years cycles. Since 2009 HOFOR has been installing permanent data loggers in Copenhagen, with 185 installed between 2009 and 2014. The loggers are connected to the internet via Leif Koch's (almosleak.com) ALMOS LEAK (Acoustic Leak Monitoring Online System), using Google Maps to highlight the status of pipes near a logger on their dedicated website. Status can be blue (normal), red (leak) or yellow (needs further investigation) with leaks being searched for and examined every night. To minimise the impact of the loggers, since 2012, they are now located underground in dedicated Spider Logger units. Before 2014, HOFOR typically found 600–700 leaks a year. In 2014, 425 leaks were identified, indicating that small leaks which were previously only detected during the triennial inspection are now being identified more rapidly.

5.8.2 Case Study 5.2: Data Logging and Network Optimisation

Excess water pressure within a distribution network increases the loss of water through the system as well as increasing energy consumption. Data loggers are used to monitor service reservoir, upstream and downstream pressure (inlet logger) and for critical pressure points within the network such as pressure regulating valves. Booster and distribution pump pressure can be optimised (removing unused excess network pressure) automatically or be remotely manipulated. Information from the loggers is integrated into a utility's smart water network. Data alerts are displayed for network breaches, open boundary valves, flow meter failure, sticking pressure-regulating valves, and pipe bursts, along with monitoring the utility's response to these and how they have been resolved.

i2O (i2Owater.com) has developed a suite of data loggers and supporting services for collecting, sending (iNet and DNet) and receiving data and sending the data to the utility, and allows the remote optimisation of data usage (oNet) and how the data is acted on. This is a network-wide water optimisation system operating on a single platform for monitoring water flows and network pressure.

New loggers can be added to the network where needed and the extant loggers can in turn perform new services via firmware updates and upgrades. Implementation is made as simple as possible, without compromising network security. The addition of additional loggers and services is intended to add increased functionality and thus more and denser data generation as and when the utility wishes to take this on.

Four examples of the i2O approach in three countries are outlined below:

1) **South East Water (UK, southeastwater.co.uk):** The system was used in 200 DMAs, with savings coming from 118 systems. 7,200 m^3 of water was saved per day or 2.628 million m^3 per annum which with a metered tariff of £1.63 per m^3 is the equivalent of £4.28 million pa.
2) **Syabas, Malaysia (syabas.com.mk):** Pressure regulating valve optimisation was carried out in 70% of the utility's district meter areas, through 700 i2O systems. This resulted in 95,000 m^3 per day in savings and a 48% burst rate reduction. Overall savings of £7 million per annum equate to £10,000 per system each year.
3) **Anglian Water (UK):** A pump automation optimisation programme resulted in a 16% average reduction in pressure, resulting in 8.4% energy savings. In addition, 31 properties were removed from the low pressure register by ensuring they had adequate pressure for the first time. There was a 38% reduction in the predicted burst rate and a 55% reduction in detection hours needed. This project had a return in investment of 7.4 months.
4) **Manila Water (Philippines, manilawater.com):** Smart pressure management was used for a single DMA zone. This resulted in 283 kWh per day reduction in energy costs or £12,866 per day, along with satisfactory customer service pressure improving from 88% to 99.8% and a 30% reduction in management resources for monitoring the zone, saving a further £20,000 pa.

Overall, as of the start of 2014, 6,000 data logging devices were installed in 66 utilities in 22 countries, saving 0.25 million m^3 of water a day (91.3 million m^3 per annum), or 41 m^3 of water saved per device per day. Operational benefits seen to date include a 20% average leakage reduction and in consequence, fewer customer complaints. Lower water consumption in turn reduces energy used by 20% due to less pumping and treatment. Pressure management lowers burst frequency by 40% with a consequent five-year increase in asset lifespan. The return on investment is typically within 6–18 months.

Principal source: Savic (2014).

5.8.3 Case Study 5.3: Developing a Leak Detection and Management System in Jerusalem

Hagihon (hagihon.co.il) was established in 1996 and provides water and sewerage services to approximately one million people in Jerusalem and the surrounding townships in Israel. Non-revenue water is 10.5% overall, 3% in the more recently developed sections of the city. The utility has 105 DMAs and new ones are being added as the utility expands its coverage.

Traditionally, leaks have to be reported before that can be repaired; what cannot be seen cannot be acted upon. At Hagihon, this is now carried out through a call centre for the continual reporting of leaks and bursts which in turn alert quick-response crews to repair the leaks as soon as possible resulting in sudden water cut-offs to consumers, which usually take place without warning, due to the need to repair the pipe burst as an emergency response. Customers suffer as they lose their water supplies and do not know when they will be reconnected.

Before the leak is observed, enough water can be discharged to damage surrounding infrastructure, and in time, properties. The repair process, from surface lifting and digging to refilling and re-surfacing can be a further cause of damage, especially to other utility assets and roads. The continuous on-line analytics of flow, pressure and water quality parameters at the district meter area level enables the detection of anomalies and alerting the monitoring system about these. This monitoring network includes permanently installed acoustic sensors for the early detection of hidden leaks.

The more notice maintenance teams have, the more effectively they can carry out leak repair work enabling them to draw up work schedules and then prioritise tasks by location (proximity to other pending tasks) and urgency. Here leak management is carried out through the following steps: [1] Following the leak alert, noting, locating and planning the intended repair and setting a time for it, and allocating this to a team within their schedule. [2] Informing authorities so as to minimise traffic disruption and locating and marking all other utility and allied infrastructure in the vicinity. [3] Confirm that the precise location of the leak has been determined, to minimise repair time, cost and disruption. [4] Alert customers about the repair in good time to ensure that they are informed and can respond if necessary. [5] Relate all planned repair work to other, low priority small leaks in the vicinity to see if any of these can be repaired at the same time. [6] Inform customers where leaks in their private networks have also been identified and how these can be addressed.

The smart elements have been developed in three stages:

1) **Real-time network monitoring:** The smart meter provider TaKaDu collects and processes data from Hagihon's SCADA network, which is analysed by the system in real-time. This provides a basis for 'normal' network performance at any point in time. Alerts occur when the system deviates from normal, to maximise the warning time at the DMA level for potential water quality issues and other problems which may result from the leak.

2) **Acoustic monitoring of the water network:** The Aquarius-Spectrum (aquarius-spectrum.com) fixed network wireless acoustic sensors provide a real-time graphical representation of every point of failure, including history and statistics. This service currently covers nearly half the water distribution network and is used to optimise and synchronise maintenance activities and generates daily updated maps of active leaks.

3) **In-pipe trenchless repair technology:** A trenchless (no dig) leak repair is provided by Curapipe (curapipe.com), detecting and sealing leakage from within the pipe.

Currently, Curapipe and Hagihon have approval of the Israeli Ministry of Health for the trenchless repair of small (4–8") pipelines. The technology is anticipated to be rolled out to larger diameter pipes in the future.

In order to continually monitor the water network for leaks, acoustic sensors are placed every 200–400 meters and activated every night for 10 seconds, with data sent via a mobile phone connection. The data generated is then analysed to build up a statistical probability of a leak, assessing which anomalies are in fact leaks, and locating the leak through the correlation of data from adjacent sensors. An assessment of the system's current leak status is generated each morning, using Aquarius software to alert and locate leaks and to add these as a dynamic layer in Hagihon's GIS system, pointing out any active leaks. The GIS system also allows the identification and recording of other network problems, including noise-making meters, non-return, valves and blocked pipes, so that these can be incorporated into the maintenance schedule.

Full-scale sensor installation commenced in October 2014. By April 2015, there were 1,700 sensors installed (out of 2,700 planned), covering 510 km of a 900 km network. During this time, 80 hidden leaks were reported and 52 repaired, along with 22 private-network leaks being identified and 60 malfunctioning devices revealed and replaced. During the first six months (October 2014 to March 2015) NRW fell from 13.5% to 10.5% and over 100 leaks were identified and repaired before customers knew about them. In effect, this means that customers are not aware of the pipe breakage having occurred. In the best performing DMA, a 6% NRW was reduced to 3%. Operating and maintenance costs were lowered even though the standard of repairs improved due to more efficient staff deployment. Collateral damage from leakage also fell due to the shorter time before the leak was repaired.

Principal sources: Yinon (2015a); Yinon (2015b).

5.8.4   Case Study 5.4: 'Mapping the Underground' for Locating Utility Assets

In the UK, four million holes are dug every year by utilities, one million in London alone (Parker, 2013). Each hole may impact other utility assets, and so there is a need to accurately locate each asset and asset failure as well as the presence of other assets. Along with 396,000 km of water mains and 353,000 of sewers, there are 275,000 km of gas mains, 482,000 km of electricity cables and an estimated 2 million km of fibre optic cables (Parker, 2013).

The extent of utility systems and the need to dig up pavements and roads to repair one or more of them is regarded as increasingly unacceptable in urban areas, even though their services are desired. The 'Mapping the Underground' project seeks to develop an integrated means of appraising and locating these assets. In 2005–2008 work concentrated on means of locating, mapping, data integration, asset tags and networks. In 2008–2012, this moved to developing multi-sensor devices and asset assessment protocols. The second phase (2012–16) sought to develop multi-sensor devices for the remote sensing of asset condition (Parker, 2014).

This involves locating, mapping in 3-D and recording, using a single shared platform: each asset's location and position to create a digital record. The current phase is under trial with Severn Trent (stwater.co.uk), Affinity Water (affinitywater.co.uk) and South Staffs Water (south-staffs-water.co.uk). With the information about sub-surface assets being transferred above the surface, a realistic picture of these assets can be developed for the first time. This may enable more effectively targeted interventions, using fewer and smaller holes, along with lower costs and collateral damage.

5.8.5   Case Study 5.5: Energy Efficient Pumping in Spain and Brazil

The Tarragona Water Consortium (ematsa.cat) serves 700,000 people in Spain. The utility has 23 pumping stations which use 56 GWh of energy a year, accounting for 20% of their total operating spending. By moving to larger regulation tanks (from 50,000 m^3 in 2005 to 175,000 m^3 in 2008) the utility had greater flexibility to concentrate pumping during off peak energy periods as well as using a more flexible array of small and large pumps. Between 2005 and 2008, the proportion of energy used during off peak hours rose from 61% to 73%.

The next priority was to minimise the energy used in backwashing sand filters through monitoring filter plugging detection. This reduced the amount of water needed for backwashing each year from 250,000–300,000 m^3 in 2010–11 to 150,000 m^3 in 2012–13, with a 40% reduction in related energy costs.

These general approaches were subsequently optimised by adopting real-time pump management and control to ensure the most effective combination of pumps being used as well as the best time for pumping and to modify pumping on actual rather than assumed demand. Derteco's Aquadapt (see Chapter 3.5.1) was deployed to further integrate and optimise these processes. As well as automating the control of pumping time and pump deployment, the Aquadapt system blends the various water sources available to the utility to ensure that they make the greatest use of gravity-fed (and less energy-intensive) supplies in relation to demand. The software has been developed to deliver these in real-time. A pilot test in one pump station, responsible for 30% of energy consumption was carried out in 2013 using in-house automation. In 2014, this was extended to pumps covering 80% of energy consumption. These operations were managed using Aquadapt in 2015. Despite rising energy tariffs, energy costs were reduced by more than 10% (Fargas-Marques, 2015).

Aegea (aegea.com.br) is the water utility division of Grupo Equipav (grupoequipav .com.br). Water shortages across Brazil in recent years have resulted in restrictions on water supplies in some states as well as increased electricity tariffs, as hydroelectricity is Brazil's primary energy source. As energy accounts for 24% of Aegea's operating spending, the utility has sought to minimise energy consumption. The first phase was to identify how much energy is needed to pump water across the network. Aegea found that 0.077 kWh/m^3 was needed to move water from its source (lagoons) to the water treatment works and to the water mains, 0.926 kWh/m^3 from the trunk mains to the water network and 0.121 kWh/m^3 from the water network to their customers. Four pumping stations accounted for 78% of their energy consumption.

Improved pumping efficiency is being developed through short and long-term plans based on the deployment of IT approaches covering asset management, customer information systems, hydraulic network modelling and the comprehensive logging and analysis of system failures. Since 2010, an 18% improvement in energy efficiency has been gained: in 2010, 19.1 million m^3 of water was delivered at 1.12 kWh per m^3 of water. In 2014, 26.0 million m^3 of water was delivered at 0.95 kWh per m^3 of water (Carvalho, 2015).

5.8.6   Case Study 5.6: Smart Water in Malta – The System

Malta has a total population of 410,000, with water provided via 140,000 connections for 240,000 household premises. The island has no rivers or lakes meaning that its water resources are under some considerable pressure; 34–48% of internal renewable resources have been abstracted annually since 2005 (Eurostat). Water Services Corporation (WSC, wsc.com.mt) supplies 80,000 m^3 of water a day to its customers, using three reverse osmosis desalination plants and groundwater pumping. In the 1990s, non-revenue waster was in excess of 45% in real and apparent losses, leakage and unbilled water consumption. This was despite 65% of the 2,300 km mains network having been installed since the 1970s.

Before the smart meter roll-out (see Chapter 4.6.5 for the smart meter roll-out programme), the distribution network was rationalised into 300 zones and subzones all of which are separately metered and logged along with 214 pressure reducing valves being deployed across the system and 28 variable speed drives installed on the network pumps. As a consequence, network leakage fell from 93,000 m^3 per day in the mid-1990s to 10,800 m^3 per day in 2013. The network's ILI was over 10 in the mid-1990s, falling to 5.0 in 2004 and 2.1 in 2013 and ILI in the island of Gozo has been in the 1.4–1.8 range since 2004.

Other losses were addressed using the smart network. For example, minimising billing errors through a modern billing system, eliminating meter misreads via smart meters and physical inspections when necessary to ensure data reconciliation, meter under-registration and the theft of water. Meter under-registration has turned out to be the biggest cause of apparent losses while theft was relatively small. By the end of 2013, data collection and billing systems were substantially in place, although with the expectation that both offer substantial scope for future improvement as the quality and reliability of data improves over time.

As of the end of 2013, a network of 250 receivers had been installed to cover the islands along with 202,000 meters with transmitters. A total of 130,000 customers were being billed through remote reading. By the end of March 2016, 227,000 out of 266,000 properties (including shops and offices) had been connected and 199,000 were being billed remotely (Pace, 2016).

A web portal for customers to manage their water consumption was launched in 2015. The current emphasis is on the integration of geographical information systems (GIS), SCADA and advanced meter management (AMM) and ensuring that the district metered water zones are correctly defined. Advanced meter management has been applied as a non-revenue water and leakage management tool and to consolidate the utilities operations to make them perform in a more efficient manner. In addition, by converging bottom–up (customer metering) and top–down (district metering) approaches, a proper water balance is being obtained along with identifying apparent loss management. Accurate consumption profiles are now being developed over time, including seasonal water use data, which also enables the utility to establish and quantify the qualitative gains that smart water metering and management can offer, thereby increasing customer satisfaction and encouraging them to be more supportive towards innovations such as smart metering. From here, data can become a tool to allow the utility to develop accurate customer segmentation by their needs and usage. Billing anomalies and illegal consumption are the next targets.

A universal smart meter roll-out on this scale is something of a learning exercise. The integration of GIS and AMM will in time determine the optimal location of smart grid components at the receiver layer. GIS is being used for data pattern analysis across the island. Real-time data is also providing a more accurate picture of how meters degrade over time and this will enable meter replacement cycles to be optimised, based on real rather than assumed data.

As in many other cases, customers remain far more concerned about their electricity costs than water costs and the linkage between water and electricity consumption provided by the multi-utility metering approach is expected to reduce water consumption over time. Customer support is gained in stages; they may not openly appreciate projects such as smart meter networks, but there is a growing appreciation that Malta's circumstances made it necessary to adopt strategies such as this (Pace, 2016).

5.8.7 Case Study 5.7: Wireless Enabled Sewerage Monitoring and Management

Anglian Water (anglianwater.co.uk) is based in eastern England, serving an area characterised by limited water resources, a flat landscape and a rapidly rising population. Sewernet is an Anglian Water project based on the effective integration of a sensor network linked to information analysis, management and control, and generating the appropriate level of visual analytics for operators. The Sewernet operates as an intelligent sewer network. Sensors are deployed across the sewerage network, providing data to the operating information management system. Within the network, each sensor unit has its own data management processes along with peer-to-peer data analysis and comparison between the sensors within the network.

The utility can link the various elements of a sewerage network and sewage treatment system through internal and external data. For example, flow, power and capacity of sewage at the pumping station, for storm retention tanks and combined sewer outflows both before and after the sewage treatment works, level, capacity and flow, volume, quality, power and chemical usage at the sewage treatment works, and quality and flow at consented outfalls. These are then related to current and forecast weather, and water flows in the allied distribution network.

Anglian Water has been using telemetry based sewage network monitoring since 2009. The current emphasis is on developing capabilities for the real-time monitoring of asset condition and to be able to predict asset failure, along with risk-based prioritisation using real-time data and the systematic measuring of sewage flows. The longer-term aim is for a self-maintaining and fully automated sewage system, that requires a minimal input from operators along with linking asset operation to short and longer term weather patterns and the ability to manage non-sewered assets, such as those in more isolated communities.

Sewernet provides data for the utility managers' alert dashboard, highlighting areas for attention and to generate data feedback (a systems intelligence engine) for the optimisation of assets in the context of system capability and loading, including assisting in the efficient detection and removal of sewer blockages.

Inferred information generated by the alarm management system includes the potential for pollution incidents, flooding (internal and external flooding as well as localised flooding), partial and complete sewer blockages, siltation and screen blockage, infiltration into the network, CSO and emergency overflow (EO) operation and pump inefficiency. This data is in turn fed into catchment models (rainfall sensitivity, capacity of the catchment, priority locations for potential overflows and so on, pumping capacity and pipeline constraints and the condition and maintenance status of the assets) and applying the integrated data to current and predicted flows. Flow management involves integrating data from weather radars, actual rainfall, the storm sewer system and drainage basins. It also includes variables such as run-off characteristics, strata permeability, time and entry of flows, soil saturation (soil wetness index), and current and anticipated storm sewer loading.

Principal sources: Kaye (2013); Kaye (2015).

5.8.8 Case Study 5.8: Monitoring for Sewer Overflows

In England and Wales, Detectronic's (detectronic.org) ultrasonic flow meter smart system for the Enterprise Data Modelling (EDM, a graphical overview of a system's performance) of storm sewers was accepted by Ofwat for AMP6 (2015–20). Data from rain gauges, MSFM Lite and MSFM S2 (Detectronic's ultrasonic CSO level monitors) are transmitted via 3G or GPRS to a Detectronic data collection and distribution node. Data goes to a local server where it is blended with external weather data, sewer asset data and historic data and analysed via DetecData Plus. It can also go straight to a remote client unit via DetecData Pro. Data from other data systems and legacy assets can be combined with data from the data collection and distribution node and through a client's SCADA, and made available for client analysis. This allows old data collected from what are currently obsolete or inoperative monitors to be included to provide historic benchmarking data. IBM is seeking to automate these processes via statistical tools that enable prediction modelling, scenario analysis and informed decision-making.

Full circle monitoring integrates all the data being collected, its interpretation and presentation in a coherent manner, ensuring that no data is lost or misplaced, with an emphasis on extracting, analysing and presenting the data that is actually needed. This also means that a single point within the utility is responsible for owning the data and how it is acted on. SCADA systems are prone to generate too many false alarms; client trials in the UK during 2014 found 30.7 confirmed predictions per 100 monitors using full circle monitoring against 11.2 for other approaches and 9.4 pollution incidents detected per 100 monitors against 3.9 for other approaches. In total, 160 pollution or flood preventing interventions including 105 blockages dealt with, saved the utilities £5 million in fines, along with reputational damage and post-event remedial works; for the fines alone this was 22 times the £225,000 investment.

Principal source: Woods (2015).

5.8.9 Case Study 5.9: Flood Warnings and Event Management

The Eastney catchment area covers Portsea Island and an adjoining part of Portsmouth in southern England, and sewerage services are managed by Southern Water. It has had a number of flooding incidents in recent years. The first, a '1 in 100 year' event on 15th September 2000 resulted in more than 300 properties suffering from internal flooding and over 530 had external flooding of up to 1.5 meters. The second, on 22nd August 2010, was a '1 in 16 year' event, when water levels rose by 4.5 meters in four minutes and by 8.6 meters in 40 minutes. An intelligent sewer strategy was devised in order to secure future resilience.

The project was managed by Innovyze (analytical software, innovyze.com), Southern Water and 4D (a joint venture between Veolia Water, Costain and MWH). A four-year plan was drawn up, covering AMP5 (2010–15). Project requirements were prepared in 2011, with a FloodWorks prototype operated during 2012 to demonstrate its viability to Southern Water. It was moved to ICMLive, a new platform in 2013 and the first phase was rolled out in April 2014. Final implementation was completed in March 2015.

Real-time data covering radar telemetry (Nowcast, very short range weather forecasts from the UK Met Office), rain gauges (five OTT Pluvios gauges, ott.com), water levels (nine sewer level monitors on constant operation) and pump operation (four stations) are fed into ICMLive under the InfoWorks integrated catchment management (ICM, InfoWorks and Innovyze) model. The model generates operational forecasts which trigger warnings via e-mail alerts and visual information. The system provides advance warning when pumping is needed and alerts if forecasted operations are not carried out. Alerts are also triggered if there are missing data inputs or divergence from the model occurs, allowing the operators to have a broad appreciation about operating conditions.

Principal source: Cockcroft (2015).

5.8.10   Case Study 5.10: Sewerage Monitoring in a Remote Community

Eco Center AG (eco-centre.it) manages the sewer networks for 323,000 people in 58 towns and villages in northern Italy's South Tyrol. Network monitoring started in 2001, with 40 measuring stations. Flow data from three networks is sent via GPRS (Mydatanet) to a central server at the sewage treatment works in Bolzano. The contract covers real-time sewer monitoring, monitoring station maintenance and customer billing. When the service started, it was based on dial up (narrowband) web access and now data-communications are supported to 3G or 4G. The system currently covers 21 sewage treatment works and flow meters installed in 71 measuring stations.

Due to the remote nature of the network, the robustness of the system was the main consideration at the outset. The system was designed to be maintenance free and is solar powered and remotely web enabled. It is designed to be easily upgradable, for example for a new SCADA system. The measuring stations are audited every 12 months by an electromagnetic flowmeter in the same channel. Non-contact radar is used for reading the sewerage level and velocity. Flowbru catchment analysis (flowbru.be) is used for monitoring sewer flow, surface water and rainfall for EU compliance data and is made available via a web based display for stakeholders.

The system was not cheap to install, but after 14 years of operation, it has proven to be reliable with no maintenance needed during this time. In consequence, this has been of a longstanding benefit in reducing operating costs.

Principal source: Davis (2015).

5.8.11   Case Study 5.11: Flood Monitoring and Management in Bordeaux

Greater Bordeaux in France covers 27 municipalities with 740,000 inhabitants with 56,000 hectares of urban land and 13,500 hectares of land lying below the high water level of the Garonne. The sewer network consists of 2,535 km of foul water sewers, 1,365 km of storm sewerage and 780 km of combined sewers. In 1982, two major flood events in three days prompted the city to consider developing a flood warning and management system. This initially involved an additional 1,900 km of storm sewerage, linked to 46 storage reservoirs and 49 pumping stations at a cost of €900 million or €1,200 per capita. This is a classic 'hard' engineering approach, concentrating on handling floodwaters. Over 400 'soft' measures were also implemented, designed to maximise the urban area's absorptive capacity (Erk, 2015).

Suez was asked to develop a storm water management system for the city in 1992. The RAMSES INFLUX™ storm water management system was developed to maximise the warning the city would get for impending flood events and the capacity to respond to this. This integrates weather data from a local weather radar (providing a detailed one hour forecast that is updated every five minutes) with network data (water level and flow and network systems status) along with current and forecast data for the rivers and streams in the catchment area via 41 rain gauges and 378 sensors providing 4,000 data inputs for the overall water status and management model (Erk, 2015). The predictive monitoring allows the manager to respond to potential flood events in a variety of ways including the rapid discharge of water-held facilities which may soon need to hold new water inflows, treating floodwaters (stormwater carries a heavy loading of contaminants) and flood prevention (reducing the flow of water from the catchment area by up to 45%, storing 1 million m^3 of water and the discharge of up to 100 million m^3 of wastewater and stormwater). Warning of specific future flood events can be given with at least six hours' notice during wet weather and 24 hours during dry weather. In 2016, the RAMSES (Regulation of Sanitation by Measures and Supervision of Equipment and Treatment Plants) INFLUX™ system was integrated with Suez's other flood management systems as AQUADVANCED Urban Drainage and integrated into the Suez Aquadvanced family of smart water management systems (Suez, 2016).

Bordeaux has not been affected by flood events since 1990, including during a storm in July 2013, which was more severe than the 1982 events, with 7 cm of rainfall in 40 minutes. The system is now being used in 20 cities, including three other major cities in France (Greater Paris, Marseille and Saint-Etienne), along with Casablanca (LYDEC Morocco), Barcelona (Spain), Singapore and the Yuelai eco-district of Chongqing one of 30 'sponge-cities' in China (Suez, 2017).

Conclusions

Smart water and management has been developed in response to meeting higher customer expectations and demand at a time when water budgets and supplies are becoming increasingly constrained. Three broad responses have been considered in this chapter: optimising the efficiency of the network in terms of the service delivered and the asset intensity needed; ensuring that assets continue to operate effectively and can be rehabilitated or upgraded when necessary until they have reached the end of their useful lives and; using extant assets and information about future needs so that new (additional) assets are deployed only when and where they are actually, rather than theoretically required.

While the smart domestic water meter on its own is not 'smart' per se, it is an integral component of a smart network, enabling real-time flow and loss monitoring at real-time within the DMA. In Chapter 4, it was shown that water smart metering is starting to be widely deployed. In contrast, domestic smart sewer meters are currently restricted to a single trial.

Examples of broad adoption have been seen within water and sewerage networks as well as for their overall monitoring and management. At the network level, smart sewerage is starting to be adopted at a scale comparable with that of the water network. The financial and reputational damage caused by sewage and sewerage incidents is a particular incentive here.

Optimising water and wastewater management is to some degree dependent on the effective integration of different aspects of network monitoring and management. The current emphasis for smart development and deployment lies in moving from single outcome approaches (metering water flow within a network for example) to integrating this information with other data sources (pressure within the network and leakage identification and status) and in the case of sewerage, this is being integrated with external data including current and forecast weather conditions, treatment capacity and effluent composition.

Flood management and monitoring systems have been established in a number of cities and in the case of Bordeaux (Case Study 5.11) and Portsea (Case Study 5.9) have provided a greater degree of resilience at the catchment level. New approaches are concentrating on the more focused use of local data for identifying vulnerabilities and potential alleviation strategies at the property level along with developing a detailed understanding about the most cost-effective locations for implementing a sustainable urban drainage across a catchment system through the effective analysis of large amounts of highly localised data.

The rate of innovation does not appear to be slowing. During the writing of this book, the potential scope for smart interventions has significantly grown and new approaches are continuing to emerge. For example, by remotely measuring the temperature of sewage as it flows through a network in real-time, it is possible to see how much groundwater (broken pipes) and stormwater (unmapped cross connections) infiltration is taking place as these are significantly cooler (Brockett, 2017).

There remains an appreciable difference between being able to successfully demonstrate a smart innovation at a single trial and being able to commercialise this or to see its broad adoption. A number of promising approaches ranging from a warning system to minimise water flushing when an urban sewer network is close to capacity, to DNA-based approaches for effluent identification, have failed to be realised during this same period of time. The attrition rate will remain high and there is no guarantee that the difference between success and failure will lie in an innovation's inherent merits.

In Chapter 6, the potential for smart applications tailored towards developing economies will be considered. Chapter 7 will explore ways of improving irrigation efficiency, which is of fundamental importance for water management, given the increasing competition for water resources. Finally, in Chapter 10, the potential role and extent of smart water offerings in a 'smart' and 'digital' world driven by the 'Internet of Things' will be considered.

References

Al-Musallam L B A (2007) Urban Water Sector Restructuring in Saudi Arabia. Presentation at the GWI Conference, Barcelona, Spain, April 2007.

Ardakanian R and Martin-Bordes J L (2009) Proceedings of International Workshop on Drinking Water Reduction: Developing Capacity for Applying Solutions, UN Campus, Bonn 3-5 September 2008. UNW-DPC, Publication 1, Bonn, Germany.

Baker M (2016) Taking quality water from source to tap. WWT 59 (10): 8.

Beal C and Flynn J (2014) The 2014 Review of Smart metering and Intelligent Water Networks in Australia and New Zealand. Water Services Association of Australia.

Boxall J (2016) Monitoring and Modelling Water Quality in Distribution Networks. Potable Water Networks: Smart Networks, CIWEM, 25th February 2016, London, UK.

Bracken M and Benner R (2016) Business case for smart leak detection: trunk mains. Presentation to 'Accelerating SMART Water', SWAN Conference, 5–6th April 2016, London, UK.

Brockett J (2017) Industry leader: Professor Dragan Savic, Exeter University. WWT 60 (2): 14–15.

Bunn S (2015) What is the energy savings potential in a water distribution system? A closer look at pumps. SWAN Forum 2015 Smart Water: The time is now! London, 29–30th April 2015.

Carvalho W (2015) Drought Conditions Force Water Utilities Efficiency. SWAN Forum 2015 Smart Water: The time is now! London, 29–30th April 2015.

Cockcroft J (2015) Managing Storm Water Flows Using the Eastney Early Warning System. Presentation to 'The Value of Intelligence in the Wastewater Network', CIWEM, London, 18th February 2015.

Common D (2016) Sewer robots sampling human waste may track drugs, disease through cities. CBC, 27th May 2016.

CRED/UNISDR (2016) The Human Cost of Weather Related Disasters, 1995–2015. United Nations, Geneva, Switzerland.

Danilenko A, van den Berg C, Macheve B and Moffitt L J (2014) The IBNET Water Supply and Sanitation Blue Book 2014. World Bank, Washington, DC, USA.

Davis M (2015) Real-time and near Real-time remote sewer network monitoring. Presentation to 'The Value of Intelligence in the Wastewater Network', CIWEM, London, 18th February 2015.

Dingle A (2016) The Flint Water Crisis: What's really going on? Chem Matters, American Chemical Society, December 2016, 5–8.

Dinkloh L (2016) Smart disinfection - using the validated UV dose concept. Presentation to 'Accelerating SMART Water', SWAN Conference, 5–6th April 2016, London, UK.

Dunning J (2015) Maximising Asset Lifetimes. SWAN Forum 2015 Smart Water: The time is now! London, 29–30th April 2015.

EEA (2009) Water resources across Europe – confronting water scarcity and drought. EEA 2/2009, European Environment Agency, Copenhagen, Denmark.

Erk T (2015) Smart Storm Water management in the context of Bordeaux. Presentation given to the IWA Busan Global Forum, 2–3rd September 2015, Busan Korea.

Fargas-Marques A (2015) Smart Engineering for Energy Savings. SWAN Forum 2015 Smart Water: The time is now! London, 29–30th April 2015.

Fisher S (2016) Addressing the water leakage challenge in Copenhagen. WWi, June–July 2016, P32–33.

Freyburg T (2017) Keeping Berlin ahead of the curve. Water and Wastewater International 32 (2): 8–9.

GWI (2016a) Inventing the Uber for water. Global Water Intelligence (17) 11: 59–64.

GWI (2016b) Chart of the month: Digital water savings for utilities. Global Water Intelligence (17) 12: 5.

Haeck M (2016) Standardized real-time control: Enhancing water quality and minimising compliance risk. Presentation to 'Accelerating SMART Water', SWAN Conference, 5–6th April 2016, London, UK.

Harrison J (2015) Predictive analysis in the sewerage network. Presentation to 'The Value of Intelligence in the Wastewater Network', CIWEM, London, 18th February 2015.

Hays K (2017) To DMA or Not to DMA? That is the Smart Water Question. Bluefield Research, Boston, MA.

Hinkel J, et al. (2014) Coastal flood damage and adaptation costs under 21st century sea-level rise. PNAS 111 (9): 3292–3297.

Hulme P (2014) CSO Monitoring and The Environment Agency. Presentation to 'Sewer Systems for the 21st Century.' Sensors For Water Interest Group, University of Sheffield, Sheffield, 25th June 2014.

Hulme P (2015) The Need for EDM. Presentation to 'The Value of Intelligence in the Wastewater Network', CIWEM, London, 18th February 2015.

Indepen (2014) Discussion paper on the potential for catchment services in England. Indepen, London, UK.

Innovyze (2010a) Real-time control: two cities, two stories. The impact of network simplification on property flood risk predictions. Innovyze case study, 17th November 2008.

Innovyze (2010b) The impact of network simplification on property flood risk predictions. Innovyze case study, 10th September 2010.

Innovyze (2010c) InfoWorks CS enables large CAPEX savings in Melbourne. Innovyze case study, 22nd October 2009.

Jha A K, Bloch R and Lamond J (2011) Cities and Flooding. A Guide to Integrated Urban Flood Risk Management for the 21st Century. World Bank/GFDRR, Washington DC, USA.

Jongman B, Hochrainer-Stigler S, et al. (2014) Increasing stress on disaster-risk finance due to large floods. Nature Climate Change 4: 264–268.

Johnson S B (2006) The Ghost Map: The Story of London's Most Terrifying Epidemic – and How it Changed Science, Cities and the Modern World. Allen Lane, London, UK.

Jung B S, Boulos P F and Wood D J (2007) Pitfalls of water distribution, model skeletonization for surge analysis. Journal AWWA 99 (12): 87–98.

Kaye S (2013) Sewernet. The opportunities and challenges of a wireless enabled business. Presentation to the Cambridge Wireless Special Interest Group, 22nd January 2013. Cambridge, UK.

Kaye S (2015) Intelligent Wastewater Networks. Presentation to 'The Value of Intelligence in the Wastewater Network', CIWEM, London, 18th February 2015.

Keeting K, et al. (2015) Cost estimation for SUDS – summary of evidence. Report – SC080039/R9, JBA Consulting for the Environment Agency, Bristol, UK.

Kostich M S, Batt A L and Lazorchak J M (2014) Concentrations of prioritized pharmaceuticals in effluents from 50 large wastewater treatment plants in the US and implications for risk estimation. Environmental Pollution 184: 354–359.

Lambert A O, et al. (2001) Water Losses Management and Techniques. Paper presented to the IWA Congress, Berlin, Germany, October 2008.

Lanfranchi E A (2016) Improving water use efficiency using innovative technologies: MM (Milan) experience. Presentation to 'Accelerating SMART Water', SWAN Conference, 5–6th April 2016, London, UK.

Liu J L and Wong M H (2013) Pharmaceuticals and personal care products (PPCPs): a review on environmental contamination in China. Environment International 59: 208–224.

Luu Q-H, Tkalich P, Choo H K, Wang J and Thompson B (2016) A storm surge forecasting system for the Singapore Strait. Smart Water 1: 2.

Merks C, Shepherd M, Fantozi M and Lambert A (2017) NRW as % of System Input Volume just doesn't work! Presentation to the IWA Efficient Urban Water Management Specialist Group, Bath, UK, 18–20th June 2017.

Miya (2015) NRW Reduction project saves water and connects 2.6 million additional people. Press Release, Miya.

OECD (2008) OECD Environmental Performance Reviews, Denmark. OECD, Paris, France.

Ofwat (2017) Delivering Water 2020: Consulting on our methodology for the 2019 price review. Ofwat, Birmingham, UK.

Owen R and Jobling S (2012) The hidden cost of flexible fertility. Nature 485: 441.

Pace L (2014) Managing non-revenue water in Malta: Going towards integrated solutions. SMi, Smart Water Systems Conference, London, 28–29th April 2014.

Pace L (2016) Smart metering in Malta. Presentation at Accelerating SMART Water, SWAN Forum, London, 5–6th April 2016.

Parker J (2013) The 'Mapping the Underground' project– industry/academic co-operation delivers dramatic new developments for water networks. Water Asset Management International 9.2: 11–14.

Parker J (2014) The development of an efficient water supply system (you can't work efficiently blindfold). SMi, Smart Water Systems Conference, London, 28–29th April 2014.

Pedersen J B and Klee P (2013) Meeting an increasing demand for water by reducing urban water loss – Reducing Non-Revenue Water in water distribution. The Rethink Water network and Danish Water Forum White Papers, Copenhagen, Denmark.

Ratti C, Turgeman T and Alm E (2014) Smart toilets and sewer sensors are coming. Wired, March 2014.

Reschefski L, et al. (2015) 2015 water in figures. DANVA, Skanderborg, Denmark.

Riis M (2015) Energy savings potential in a water distribution system. SWAN Forum 2015 Smart Water: The time is now! London, 29–30th April 2015.

Romano M, Woodward K and Kapelan Z (2016) Analytics for locating bursts in water distribution systems. Presentation to 'Accelerating SMART Water', SWAN Conference, 5–6th April 2016, London, UK.

Savic D (2014) Smart Water Networks: The European Perspective. CIWEM, London, 4th December 2014.

Skytte O (2016) Greater Copenhagen Utility. Presentation to 8th Global Leakage Summit, 27–28th September 2016, London, UK.

Slater A (2014) Smart Water Systems - Using the Network. Presentation to the SMi, Smart Water Systems Conference, London, 28–29th April 2014.

Sodeck D S (2016) Water loss with information overload. Presentation to Accelerating SMART Water, SWAN Conference, 5–6th April 2016, London, UK.

Stephens I (2003) Regulating economic levels of leakage in England and Wales. Presentation to the World Bank World Water Week, Washington DC, USA, 4–6 March 2003.

Suez (2016) Suez launches AQUAADVANCED™ urban drainage, an innovative solution to optimise the performance of sewer and stormwater networks and preserve the natural environment. Press release, Suez, 11th July 2016, Suez, Paris, France.

Suez (2017) Suez wins the contract for the optimisation of sewer and stormwater systems in the city of Chongqing (China). Press release, Suez, 17th February 2017, Suez, Paris, France.

Symmonds G (2015) The Challenge of Transitioning from AMR to AMI. Presentation given at the SWAN Forum 2015 Smart Water: The time is now! London, 29–30th April 2015.

Todorovic Z and Breton N (2017) Anglian Water uses mapping tools for retrofit SuDS. WWT, 60 (4): 23–24.

Traub A (2106) Intelligent leak detection using acoustic sensor networks. Presentation to 'Accelerating SMART Water', SWAN Conference, 5–6th April 2016, London, UK.

Water Active (2017) Wastewater monitoring at Severn Trent. Water Active 21 (3): 9.

Wheeler N, Francis N and George A (2016) Smarter flood risk management in England: investing in resilient catchments. Green Alliance, London, UK.

Woods S (2015) Using 'Smart Network Monitoring' to Reduce Flooding and Pollution. Presentation to 'The Value of Intelligence in the Wastewater Network', CIWEM, London, 18th February 2015.

Yinon Z (2015a) Smart steps towards smart water. Presentation at the SWAN Forum 2015 Smart Water: The time is now! London, 29-30th April 2015.

Yinon Z (2015b) SMI – Hagihon Case Study. Presentation at the SMi Smart Water Systems Conference, London, April 29–30th 2015.

Zheng Y W (2015) Integrating Data-Driven Analysis with Water Network Models SWAN Forum 2015 Smart Water: The time is now! London, 29–30th April 2015.

Zpryme (2014) Smart water survey report 2014, Badger Meter, Milwaukee, USA.

Zpryme (2015) Smart water survey report 2015, Neptune Technology Group, Tallassee, USA.

Zuccato E, Chiabrando C, Castliglioni S, Bagnati R and Fanelli R (2008) Estimating community drug abuse by wastewater analysis. Environmental Health Perspectives 116: 1027–1032.

6

Appropriate Technology and Development

Introduction

A variety of smart water approaches in developing economies have been noted to date, including in Malaysia (5.3.1), the Philippines (5.3.3) and Brazil (5.3.5 and Case Study 5.5). These have taken place where a service has been adopted by utilities in a variety of economies. This chapter will concentrate on applications that have been developed with developing economy markets in mind, with particular reference of meeting the United Nations' water and sanitation Sustainable Development Goals for 2030.

6.1 Sustainable Development and Water in Developing Economies

The 2030 Sustainable Development Goals ('SDGs') seek to 'ensure availability and sustainable management of water and sanitation for all' by 2030 as outlined in Chapter 2.1.5. The important change from the 2015 Millennium Development Goals has been the adoption of 'safe' rather than 'improved' water and sanitation.

A review by the World Bank's Water Global Practice (Kolker et al., 2016) observed that 48 countries failed to meet the 'safe' (in other words, 'improved') drinking water Millennium Development Goals targets (to halve the number of people without access to 'improved' drinking water and sanitation by 2015 against a 1990 benchmark) in 2015, while sanitation was well behind even this and given that SDG6 is 'much more ambitious' than the MDGs, developing countries 'face enormous challenges'.

Hutton and Verughese (2016) developed what the author believes to be the first comprehensive attempt to quantify the cost of meeting the various goals for access to safe water and sanitation since 1980. Annual capital spending needs by category from 2015 to 2029 are summarised in Table 6.1.

Currently, $16 billion pa is spent on capital spending (Tremolet, 2017, personal communication), along with an estimated $75 billion pa on operating expenditure and $29 billion pa on maintaining extant assets. Hutton and Verughese (2016) forecast that as well as $114 billion pa needing to be spent on new water and sanitation assets, additional operating expenditure of $92 billion pa and a further $52 billion pa on maintaining these new assets will be needed between 2015 and 2029.

Smart Water Technologies and Techniques: Data Capture and Analysis for Sustainable Water Management, First Edition. David A. Lloyd Owen.
© 2018 John Wiley & Sons Ltd. Published 2018 by John Wiley & Sons Ltd.

Table 6.1 Capital expenditure needed to meet water and sanitation SGD6.

$ billion pa	Basic	Safe	SDG6
Urban water	5.4	23.3	28.7
Rural water	1.4	13.0	14.4
Urban sanitation	13.1	29.3	42.4
Rural sanitation	5.8	17.6	23.4
End outdoor defecation	3.6	–	3.6
Hygiene	2.0	–	2.0
Total	31.3	83.2	114.5

Source: Adapted from Hutton and Verughese (2016) and the author's data.

What happens if this money is not spent? Entrepreneurs step in where governments and utilities fear to tread as people are willing to pay a premium for some form of access to safe or even less unsafe water (IFC, 2009). Gasson (2017) considered how the market for alternative (non-utility) water spending would develop under a business-as-usual scenario if utilities and municipalities are not able to invest in universal access to safe water supplies. In this case, by 2030, non-utility ('coping' or discretionary spending in the absence of safe utility supplies) spending would become larger than utility spending. This is already the case in Mexico, where tap water is used for non-potable applications and bottles for drinking water. Under this scenario, utility totex (operating and capital expenditure combined) falls from 51% of domestic and commercial water spending in 2015 to 37% by 2030, meaning that a loss of confidence in utilities forces people to spend appreciably more on other water sources, not necessarily safe ones, while continuing to starve the utilities of the funds needed to improve their services and coverage.

It is unlikely that ODA (official development assistance) for safe water and sanitation projects will increase significantly from its current level of $5.7 billion pa (Winpenny et al., 2016) and generating investor interest in water and sanitation projects remains at best challenging (Kolker et al., 2016). While increased revenues will be generated where people are connected to utility services, most of this will be absorbed by operating and maintenance costs. A survey of capital spending towards SDG6 GLASS (2017) concluded that utility spending is currently growing at 5.5% per annum, which is not adequate to achieve the necessary spending to attain SDG6. As a result, more efficient ways of carrying out capital projects and for lowering operating costs are needed. Smart water can play a significant role here.

6.2   Overcoming Traditional Obstacles

Where people are served with water and sanitation, these services may not be performing in a satisfactory manner, if at all; there is a difference between assets being installed and their providing a service. Likewise, even when water is delivered, if it is trusted to be fit to drink, home treatment or alternative sources will be used (IFC, 2009; Gasson, 2017).

In India, the average water supply in 20 utilities surveyed by the Asian Development Bank was 4.3 hours per day (ADB, 2007) and subsequent progress has been limited. Intermittent supplies affect meter reading and leakage control as well as water quality. Where households have individual water tankers to provide continual supplies, these are emptied every day prior to being refilled, which drives up water consumption. In Delhi, the average connected household spends Rs 2,000 every year dealing with this intermittent supply, on point of use water treatment and tank cleaning services, 5.5 times as much as they spend on the actual supply (McIntosh, 2009).

6.2.1 Aid-Funded Rural Hand Pumps in Sub-Saharan Africa

There is little point in building hand pumps and boreholes in rural areas unless they work and continue to work. It was estimated that in the mid-2000s, 36% of 345,071 hand pumps in Sub-Saharan Africa covering 55.5 million people were non-functioning (RWSN, 2009) while 17–30% of 79,383 pumps installed in Liberia, Sierra Leone, Malawi and Tanzania were out of service within a year (Carter and Ross, 2016). One of the chief factors behind their failure is the importing of pumps by international agencies without considering their long term viability or building up any local capacity to maintain them. In 2010, there were over 150,000 abandoned pumps in Africa, representing $2.5 billion in wasted investments (van Beers, 2015).

6.2.2 Reducing Water Losses and Unbilled Water in Developing Economies

Non-revenue water (NRW) is a challenge for utilities already facing funding shortages. Viewed another way reducing NRW offers the utility more tariffs from the same amount of water that is already being put into the network. The IB-NET survey for 2010 (Danilenko et al., 2014) was considered in Chapter 5.1. With revenues of $23.96 billion in 2010 and median non-revenue water of 28%, potentially worth $6.71 billion pa was lost in these utilities alone.

To address such shortcomings, utilities need to be able to levy tariffs that cover their operating costs and fund at least the financing of new capital costs, either through full cost recovery or sustainable cost recovery, where tariffs are blended with other revenue streams such as ODA. These tariffs need to be collected in without undue delay with the revenues going to the utility rather than being lost through corruption. For hand pumps, some basic principles are needed to ensure they operate in an effective manner. This requires pumps that have been developed to be inherently robust and capable of being managed by their local communities using readily available materials.

This chapter will explore ways that smart approaches can assist in delivering these aims and to improve service delivery and quality.

6.2.3 Developing Water Pumps that are Built to Last

Sometimes, the simplest of approaches offer the most benefit. Hardware for Pump Aid's 'Elephant Pump' costs £900 with a total cost (including well digging, staff costs and fuel) in Malawi of £3,000 per installation, or £25 per person, 60% less than conventional aid-funded water pumps. Pump Aid (pumpaid.org) provides water to 500,000 people in Malawi through 4,000 Elephant Pumps (Pump Aid, 2013). In Zimbabwe, it was found that 90% of pumps were still functioning seven years after Pump Aid was ordered to leave the country.

FairWater (fairwater.org) has developed the 'Blue Pump' which is designed to be installed in abandoned pumping facilities. In 2008, 40 pumps had been installed in Africa, serving 10,000 people. The pumps are designed to operate for at least 15 years at a cost of $7.5 per person or $0.50 per annum compared with the traditional NGO approach costing $75 per person or $15 per annum for a pump that lasts fewer than five years (van Beers, 2009). By 2015, 693 Blue Pumps had been installed in eight countries.

6.3 The Impact of Mobile Telephony

Previous chapters have highlighted the importance of data collection and communication in enabling smart networks and services to be developed. Mobile data transmission plays a central role in enabling this to happen. In developing economies, the paucity of fixed wire infrastructure, especially in rural areas, redoubles the importance of having access to mobile data transmission and smart phones.

6.3.1 The Need for Access to Services and Infrastructure

The broad adoption of smart mobile devices remains a work in progress in most developing economies. It requires both an infrastructure capable of supporting large scale and high speed data transmission (mobile broadband requires 3G and above) and either a smart phone or mobile internet connection. In Table 6.2, smart phone adoption refers to the proportion of subscribers that have a smart phone rather than a traditional mobile phone.

Infrastructure development and smart phone adoption are a particular challenge in Sub-Saharan Africa and India. Subscriber penetration in India was 47% in 2015 compared with 78% in Vietnam and 66% in Indonesia and 20% of the subscriber base in India have smart phones meaning that 9% of the population of India had a smart phone at the time. In addition, 85% of connections in India were for 2G (GMSA, 2016a). In contrast, subscriber penetration was 73% in China with 68% of these being smart

Table 6.2 Mobile infrastructure and service penetration in 2016.

	3G/4G technology	Subscriber penetration	Smartphone adoption	Mobile internet
Middle East and North Africa	47%	70%	46%	36%
Sub-Saharan Africa	32%	44%	28%	28%
Asia-Pacific	53%	65%	51%	50%
Latin America	61%	70%	55%	52%
Global total	55%	65%	51%	48%

Source: Adapted from GMSA (2017).

Table 6.3 Regional coverage in Sub-Saharan Africa, 2015.

	3G/4G technology	Subscriber penetration	Smartphone adoption
West Africa	21%	47%	23%
Central Africa	11%	33%	19%
East Africa	23%	46%	17%
Southern Africa	29%	42%	24%
Sub-Saharan Africa	23%	46%	23%

Source: Adapted from GMSA (2016c).

phones (GMSA, 2016b). Coverage can be greater than subscriber numbers may indicate. For example, 98% of India by area has at least 2G coverage (ITU and Cisco, 2016). This is not just an emerging economy problem; in 1016, 93% of households in the UK were covered by all 2G services, 88% for 3G and 46% for 4G, with service providers obliged to provide voice and text coverage to 90% of the UK landmass by 2017 (Rathbone, 2016).

Service development varies within a region, as seen in Table 6.3. In Asia-Pacific, access to mobile internet services was 45% overall, but 81% in the developed countries and 37% in the developing economies (GMSA, 2016b). The mobile communications market is rapidly evolving. For example, in the Asia Pacific region, subscriber penetration is expected to rise to 76% by 2020 while in Sub-Saharan Africa, the percentage of mobile phones that will be smart phones is forecast to rise from 28% to 55% by 2020 (GMSA, 2017).

Universal access and coverage is ideal, but what matters more is that householders and monitors in remote rural areas have some access to these services and that data communications are adapted to deal with incomplete networks and service interruptions. Perfect communications are not needed for all forms of mobile-based data collection and transmission.

6.3.2 Making Innovation Matter – Mobile Money and Water

In order to contribute to the development of water assets and services in developing economies, new technology needs to satisfy at least three of the following six criteria: to be either scientifically, economically, environmentally and socially viable; to use appropriate levels of technologies for developing countries and to be able to ensure that new assets remain in working order.

As the mobile communications revolution shows, the adoption of innovation in developing economies can be a more pragmatic and adaptive process than in developed economies, due to the relative absence of prior infrastructure and services. In the case of mobile money in Africa, the continent is leapfrogging the need to expand the various bank branch networks by going directly to customer mobile platforms.

In 2016 there were 277 mobile money services being operated in 92 countries (GSMA, 2017) with 556 million registered and 174 million actively used mobile money accounts. A total of 141 million of the active accounts are in Sub-Saharan Africa and South Asia.

In 37 of these markets, agents account for 10 times as many access points as bank branches.

For example, in Kenya, there were 16.6 million active M-Pesa customers in 2016, and 101,000 agents. This means that while there are 11 ATMs and six commercial bank branches per 1000,000 adults, there are 538 M-Pesa agents.

Nairobi's City water and Sewerage Company (NCWSC) uses Safaricom's M-Pesa for mobile payments from households in the Kayole Soweto settlement and to distribute subsidies to poorer households for first-time water connections through its Maji Mashinani ('water at the grassroots') programme. The settlement has 89,000 inhabitants and as of October 2014, there were 2,217 metered accounts under the programme, serving approximately 8,970 people, with a connection fee of KS 8,215 ($80). Revenues from the connection fees have covered the cost of the project which has subsequently been scaled up to increase coverage. The customer makes a meter reading when required and sends the data by short message service texts (SMS) to NCWSC, which sends the bill by return. The bill is then paid through M-Pesa. Customers do not pay for their SMS texts, while NCWSC pays KS0.80 for the response SMS (World Bank, 2015).

In Tanzania, the Dar es Salaam Water and Sewerage Corporation (DAWASCO) started accepting mobile payments for utility bills in 2009 via Vodacom Tanzania. By 2013, DAWASCO's revenues had increased by $0.54 million through improved revenue collection. The service allows customers to pay their bills at the time of their choosing (GSMA, 2017).

In the case of India (Table 6.4) access to fixed wire telephones was always limited and having peaked at 42 million in 2006, their use has since declined. The numbers for mobile telecoms are appreciably higher than for unique subscribers as a significant number of people have at least two handsets. Even so, a disruptive invention that was originally seen as a specialist and premium service in the most advanced economies has become commonplace in markets such as India because it delivers tangible benefits for customers who are willing to pay for them. Water and sanitation under the present paradigm, have made incremental progress.

Table 6.4 Development of telecoms in India and water and sanitation, 2000–16.

Million people	2000	2006	2010	2016
Fixed wire telecoms	27	42	37	24
Mobile telecoms (accounts)	2	99	584	1,127
Total for telecoms	29	141	621	1,151
Access to household piped water	208	242	282	359
Access to 'improved' sanitation	261	323	417	513
Dependent on open defecation	656	652	603	564

Water and sanitation data for 2016 is for 2015.
Source: Adapted from TRAI (2004, 2008, 2011 and 2017) and WHO/UNICEF (2015); WHO/UNICEF (2012); WHO/UNICEF (2008).

6.4   An Overview of Smart Water Initiatives Seen in Developing Economies

The development of smart applications in developing economies is at an earlier stage than in the cases previously discussed. Over time, approaches used in developed economies will be adopted as well, as already seen in Chapter 5. This section reviews various initiatives that have been specifically developed with developing economies in mind.

A number of reviews covering the potential for smart applications for water and sanitation in developing economies have been published in recent years. McIntosh and Gebrechorkos (2014) consider the role of information and communication technologies in enabling the development of smart water management in developing economies from a variety of perspectives. Krolikowski, Fu and Hope (2013), and Nique and Smertnik (2015) review the impact of mobile payment systems on access to water and sanitation services and Sibthorpe (2016) examines the impact of mobile communications for women. Hope et al. (2011) reviewed the potential for smart water metering. Prat and Trémolet (2013) provides an overview about the development of sanitation apps and the Toilet Board Coalition (2016) considers the potential for smart approaches to sanitation and the potential obstacles to their broader adoption.

6.4.1   India's Smart Cities Mission

The smart cities mission is a R980 bn ($15 bn) project for developing 100 smart cities in India. Each city will receive central funding of $15 million pa for five years, matched by state funding, with 20 cities shortlisted for the initial phase in June 2016, covering 360 million people. By September 2016, 60 cities had been selected. 21 'smart solutions' have been identified, including three for water (smart water meters and management, leakage identification and preventative maintenance, and water quality monitoring) along with treatment of wastewater. All area-based developments will offer 'adequate' water supply along with water recycling and rainwater reuse. In the case of Pune, R28 billion has been budgeted for continual water supply, R10 billion for improving river water quality and R19 billion on other water projects for real time monitoring and management of the entire water cycle (Ongole, 2016).

6.4.2   Remote Pump Condition Monitoring

The Unlocking the Potential of Groundwater for the Poor research programme is trialling a remote device for monitoring how hand pumps operate over a period of time (Purvis, 2016). Researchers from WaterAid, Overseas Development Institute, British Geological Survey are examining data generated by 600 hand pumps in Uganda, Malawi and Ethiopia. The waterpoint monitor was developed by the Smith School of Enterprise and the Environment at Oxford University. The device has a small low-cost accelerometer which tracks the arching movement of the pump's handle to estimate water usage, sending the data via SMS to a web-based dashboard. The aim is identify how and why pumps fail and to develop effective ways of addressing this.

In Kenya, the waterpoint monitor is being used by the FundiFix service. This is managed by social entrepreneurs to identify where pumps are not in use or may be malfunctioning so that they can be repaired by local maintenance providers. Currently, 300 hand pumps serving 60,000 people are being monitored in Kwale, Kitui and Kakamega, with data sent to a central server. In consequence, the average repair time has fallen from a month to two days, with a guarantee of repairs within three days. After trials In Zambia and Kenya in 2011–13, a maintenance fee was introduced once a guaranteed service quality could be delivered. The aim is to reduce pump downtime to as close to zero as possible. In Kwale, the high-frequency 'noise' in the accelerometer data has also been used to assess groundwater levels so that users can have advance warning about groundwater resource depletion (Colchester et al., 2017).

An earlier trial using waterpoint at Kyuso (Oxford/RFL, 2014) resulted in the time that pumps were functioning rising from 67% to 98% with a reduction in repair time from 27 days to three and an increase for willingness to pay for pump maintenance due to improved pump performance from 20% before the trial to 80% at its conclusion.

6.4.3   SWEETSense – A Multi Use Monitor

The SWEETSense sensor was developed by Portland State University for monitoring the performance and flow of water pumps and has subsequently been used in a wide variety of applications (ITU and Cisco, 2016) including pump monitoring in Kenya, monitoring latrines in Bangladesh and hand washing in Indonesia. The sensors generate data continually, which is transmitted via a mobile connection to the SWEETData™ internet database for analysis. In 2016, the sensors cost $100, which is expected to fall through economies of scale.

Using a SWEETSense pump sensor adds 10% to the price of a hand pump in Rwanda while reducing down time by 80–90%. The monitors currently have a battery life of 12–18 months and 200 units have been deployed to date (ITU and Cisco, 2016). SWEETSense sensors were used in Jakarta, Indonesia to monitor hand washing behaviour in latrine blocks. In this case, it was found that actual hand washing was less frequent than the users had stated under self-reporting. This highlights how smart approaches can improve on the data generated under traditional assumptions (Thomas and Matson, 2014).

6.4.4   Data Collection, Transmission and Interpretation Systems – mWater

Access to data about water resources, where they are located and the quality and quantity of the water they can provide is one of the greatest challenges for developing economies, especially in rural areas.

This can be addressed through a people-based approach. For example, mWater (mwater.co) is a not-for-profit tech start-up founded in 2011 to enable people in developing economies to generate, disseminate and evaluate water and sanitation data from remote locations. In 2017, mWater had 10,000 users in 93 countries, with 25,000 surveys a month being received covering 350,000 public and private sites.

The mWater Surveyor mobile app is provided free for Android devices. Field data is collected using a smart phone and sent to a portal, which is a cloud based platform, allowing operators to collect and transmit survey data under all conditions, so that data is not lost when a connection is lost. The Explorer app provides a set of standard forms for mapping water sources, functionality, quality and sanitation status. These apps are designed to be adapted by groups for their purposes, based on data visualisation and the ability to overlay various data layers as needed. Results from submitted surveys are stored in Excel or CSV formats and visualised in real time. The portal is designed to be easily understood, with visualisation made as simple and effective as possible. Open-source software has been used, ensuring free access (Ross, 2016).

In Haiti, mWater has been deployed by Haiti Outreach (haitioutreach.com) to survey all water points in three of the country's ten departments between 2015 and 2016 for its functionality and potability. This data was overlaid with satellite imagery identifying every household in these areas and presented on a dashboard. This allows users to identify where a household is within five minutes (500 meters) of a water source and its functionality; working (78.6%), in need of repair (4.6%), not working (15.8%) or no longer there (0.9%), allowing water operators and NGOs to prioritise investments. The project is now being rolled out in other departments.

In Tanzania the city of Mwanza has used mWater to monitor all water sources developed by NGOs. They found that shallow wells were the most popular choice for their new water sources, but these surveys also found that 90% of these wells became contaminated within a year. As a result, NGO built shallow wells are now being refused certification and NGOs are obliged to develop more sustainable water sources. With free source monitoring at the community level, water sources can be more effectively managed.

Other applications developed to date include mapping an urban water network and its operational efficiency (Dar es Salaam, Tanzania), assessing new water points by actual need rather than political considerations (Uganda) developing common water data and sanitation reporting standards (WaterAid and water.org) and alerting farmers about contaminated water sources (Tanzania).

Another consideration is the utility provided by a service against its cost. Fisher et al. (2016) evaluated seven mobile survey tools with regards to their performance during rural water and sanitation monitoring. Here, mWater was the second best performer in terms of ease of use, reliability and service delivery. While Fulcrum, the best performer, cost an estimated $4,788 per annum for ten users, mWater is free. Fisher (2016) also noted that the performance of the five paid mobile survey tools 'was not significantly better than the aggregate scores' of the two free services (mWater and ODK) once cost was taken into account. Developing economies do not need to be premium markets.

6.4.5   Managing and Monitoring Losses

In Kenya, sensor-based tools are being tested to implement real-time monitoring systems to overcome key low income market challenges associated with non-technical losses, such as poor operations, low payment efficiency, and theft. Service providers Upande, and BRCK together with the Kericho Water and Sanitation Company (KEWASCO) in Kenya are using smart meters with alert modules and low-cost solar-powered data loggers to reduce non-revenue water losses. The data loggers measure water flow and transmit data to the cloud providing accurate data on water usage and loss using the blockchain distributed ledger data technology (Wyman and JPM, 2016).

The 'Reduce Water Leaks by a Mobile Device' app has been developed by Trịnh Quốc Anh, Nguyễn Trần Quang Khải and Võ Phi Long, at Ho Chi Minh City University of Technology. When a leak is noticed, the observer touches a dedicated app which alerts SWACO, Ho Chi Minh City's water utility about the leak and locates it using the smart phone's Global Positioning System co-ordinates. The leak is logged into SWACO's system where priority is given to the leaks with the highest number of notifications. Logging the number of notifications acts as a verification service. The intention is for as many people as possible to have access to the app to maximise the number and rapidity of notifications as well as becoming a means to increase people's awareness about water conservation in general. The app won the first prize at the nationwide Smart Water Innovation Contest 2016, organised by the Swedish Embassy in Hanoi. A fully operational version of the app is under consideration (Minh, 2016).

6.4.6 Smart Sanitation – Logistics and Lavatories

Traditionally, the cost of collecting fecal sludge has not been addressed due to the service being carried out in an ad hoc manner. Likewise, lavatory units may be serviced on a routine basis, leading to deteriorating conditions for users unless action is taken. SMS and smart phone messaging, real-time geo-locators and digitised customer relationship management (CRM) are being used to optimise both the logistics of waste collection and servicing and repairing facilities along with gaining a better appreciation about actual customer needs and preferences (Toilet Board Coalition, 2016). A wide variety of approaches have been seen for improving the effectiveness of stand-alone lavatories, two of which are discussed below.

An 'e toilet' has been developed by Eram Scientific (eramscientific.com) for use in urban and peri-urban locations in India. The units are self-cleansing, via an automatic pre-flush and periodic platform washing (a 5 litre flush after every five uses) and offer 1.5 litre short flushes and 4.5 litre full flushes as directed by a sensor. The stand-alone units have a tank holding 225 litres of water. Entry for public units is through gate after paying a fee. Each unit has a GPRS connection to clients and the company for monitoring its performance and status (water levels, usage patterns and when to service, for example) and their locations can be found via a dedicated app. 2,100 units have been installed to date, including 900 school units. The units cost R100,000 for the school model and R400,000 for the public model (Baby and Vinod, 2012; Eram, 2017).

Saraplast (3sindia.com) developed its 'Mobi-Loo' for construction sites and special events along with households, public sites and disaster relief zones. M2M and mobile service platforms are used to optimise how the units are moved from event to event, and a mobile app is used for tracking the units' toilets via geo-location and to aggregate the services for waste management (Toilet Board Coalition, 2016).

6.4.7 Sanitation Apps

Table 6.5 aims to synthesise the apps noted by Prat and Trémolet (2013) in their overview of the sanitation app developments in the wake of amongst others, 'hackathons' sponsored by the World Bank. As they noted at the time, these initiatives and others, are works in progress and demonstrate the potential of involving people not normally associated with water and sanitation services.

As with the water leak alert app noted in section 6.5.5, these give an idea of the variety of apps that are under development, some of which may become fully realised and adopted and other less so, if at all. Some may be widely and beneficially used within a relatively small area and may not attract wider notice, while others could be adopted in a number of countries. Some of this is down to circumstance; it is also a matter of developing apps that combine a compelling message and are easy to use.

Table 6.5 Sanitation apps.

Education for behavioural change	**SunClean**: to teach children about WASH.
Gaming apps as a way of informing people about the dangers of open defecation. It also aims to encourage hand washing.	**Clean Kumasi**: a platform for enabling people in an area to work towards eliminating open defecation.
Open defecation is a sensitive subject and so apps need to be developed to effectively reflect local customs, age cohorts and elements such as sense of humour.	**Loo rewards**: form hygiene teams to encourage hygienic behaviour.
	SunClean and San-Trac: to encourage children and adults to wash their hands.
Self-reporting	**Taarifa**: open source platform for enabling people to report their local circumstances.
For reporting problems with private and public facilities and to alert public authorities about problems.	**mSewage**: to identify where water sources are at risk from sewage contamination.
Depends on the willingness of people to report problems and to see that their concerns are dealt with effectively.	**mSchool**: for monitoring the condition of facilities in schools.
	See also Clean Kumasi
Mapping	**Sanitation mapper**: area based mapping for monitoring sanitation facilities.
For surveying community facilities to identify where services are needed and open defecation is taking place.	**Toilight**: to locate the nearest facility and its facilities, opening times and price.
In urban areas, GPS needs to be supplemented with ways of pinpointing a locality in a densely populated area.	**Sanitation investment tracker (SIT)**: for tracking household sanitation investment and spending.
	See also Taarifa and mSewage
Monitoring and planning	**SIT**: gather financial data to assess the overall performance of a project.
For programme managers and monitors for data collection and monitoring of costs and delivery of intended outcomes.	**Outcome tracker**: for monitoring the use of various types of sanitation facilities.
These apps need to be flexible in order to be able to be applied to a wide range of specific local circumstances.	**WASHCost calculator**: for assessing the life cycle costs of a project and to allow practitioners to adapt plans as needed.

Source: Adapted from Prat and Trémolet (2013).

6.5 Case Studies

6.5.1   Case Study 6.1: Smart Water ATMs in an Informal Settlement in Nairobi.

6.5.2   Case Study 6.2: Smart Sanitation Collection in Senegal.

6.5.3   Case Study 6.3: India – Performance-Based PPP Contract for Water Services.

6.5.1   Case Study 6.1: Smart Water ATMs in an Informal Settlement in Nairobi, Kenya

200,000 people live in Mathare, an informal settlement in Nairobi. There is no formal water provision in the area, and many people depended on water supplied by vending cartels costing Sh 50 per 20 litres ($0.50), or Sh 2,500 per m^3 ($25.00). In 2015, water vending machines were installed by a public-private partnership between Nairobi Water and Sewerage Company (NWSS, nairobiwater.co.ke), the city's water utility, and Grundfos (grundfos.com).

NWSS has laid 18 km of pipes providing potable water to automated teller machines (ATMs) which are access by pre-paid smart cards, dispensing water into a customers' container for Sh 0.50 per 20 litres ($0.25 per m^3). The ATMs are managed by community leaders, who also ensure that the pipelines are not cut by the water vendors (Wesangula, 2016).

6.5.2 Case Study 6.2: Smart Sanitation Collection in Senegal

The city of Dakar generates 1,500 tonnes of faecal sludge per day, of which 400 tonnes are uncollected. One of the constraints for extending sanitation collection is the cost. Dakar had a population of 3.14 million in 2013 (Senegal Census Data, 2013) with, officially, 41% connected to a sewerage network and 47% using pit latrines and 12% depending on unimproved sanitation. Fecal sludge collection was traditionally operated as a cartel and used to cost $150 per household per annum, equivalent to 2% of GNI per household of 3% of household consumption. For poorer households, this can be prohibitively expensive.

In 2014, ONAS, the city's water utility launched a SMS-based service enabling households to call a customer service centre when their latrines need emptying. The system was developed by Manobi (manobi.net), a Senegalese software developer, with funding from the Bill and Melinda Gates Foundation (gatesfoundation.org). Sludge collectors are alerted to each opportunity through SMS messages and bid for the business. In its first year, competitive bidding reduced the annual cost from $150 per annum to $90 and it is hoped that this will fall to $60 in time. By 2015, there were 65,000 customers, covering 0.51 million people or 15% of the city's population (Hussain, 2015; Nique and Smertnik, 2015).

6.5.3 Case Study 6.3: India – Performance-Based PPP Contract for Water Services

In December 2016, Suez was awarded a €30 million six-year contract to improve water delivery and reduce non-revenue water in the Cossipore district (population 200,000) of Kolkata. The ADB backed project aims to provide continual potable water supplies and while reducing NRW to 30%. 20% of the O&M revenues are subject to Suez meeting performance targets. 25,000 house connections and meters will be installed and Suez is using its helium gas leakage detection system, which can work effectively in areas with intermittent water supplies (Suez, 2016).

Conclusions

Delivering universal and sustainable access to safe water and sanitation to developing economies is the greatest challenge facing water management today. For human development, it is also the most rewarding. The gap between global ambitions such as SDG6 and their effective realisation is nearly as great as it was when previous initiatives were launched in 1980 and 2000. Can smart approaches make a difference here?

It is evident that there are a great variety of tools, systems and apps that are in various stages of development and deployment. Most of these have been created individually with specific regions and market applications in mind, so they will face the need to be able to interact with other smart offerings if and when they are deployed on a wider basis. Anecdotal evidence points to a degree of lassitude at the national level concerning communications operating standards which may be problematic when seeking to transpose mobile communications based initiatives from one country to another.

GSMA (2015) envisions a five layered approach towards mobile sanitation applications for developing economies. First, there is the mobile infrastructure, connecting households to the mobile telephony network. Next is the mobile operator's distribution network and mobile money agents, for the distribution and sale of household latrines by connecting households to entrepreneurs. Machine to machine (M2M) connectivity allows the householder to access a latrine emptying service in the most cost-effective manner possible. Mobile payments cut the time taken for making payments to the utility or sanitation service provider by the customer while cutting the administrative costs for the latter. Finally, mobile services allow village entrepreneurs to order hardware from the best value providers and to enable households to access service teams when their units need to be repaired.

Mobile payment systems also have a role to play in developing intermediate access to safe sanitation. Where households cannot afford to have their own sanitation facilities, entrepreneurs can develop affordable latrine units that are funded by mobile payments. Here, the aim is for household access in the longer term, as they accrue the benefits of accessing these units in the meantime. Pre-paid cards are set to play an important role in driving down the cost of payments for various services such as lavatories, obviating the cost of the SMS service and making the use of a SMS or similar communications affordable and attractive.

In countries where sanitation has been made a priority by lower levels of government, informal communication, reporting sharing of results via mobile communication has encouraged learning and adaptation, making sanitation programs more responsive and flexible. For instance, in India and Indonesia, digital communication platforms like Whatsapp were seen to be allowing junior bureaucrats to skip the traditional hierarchies and to talk to the relevant superiors, meaning that information was shared faster and more accurately and giving more motivation to produce results (Toilet Board Coalition, 2016). This is an example of the unintended consequences of a disruptive communication innovation and how it can enable new information flows to develop.

The development of smart-enabled equipment ranging from waterless, autonomous lavatory units designed for effective waste resource mobilisation (Loowat, loowat.com) to localised sludge resource recovery systems (Omniprocessor, janikibioprocessor.com) are not 'smart' in themselves and lie outside the scope of this study. What these and other innovations have is the potential to improve the impact and cost-effectiveness of the smart approaches that have been outlined in this chapter. They also have the potential to be developed into enabling mechanisms for demand management.

A study (in prep) by the author for the World Bank on capital efficiency and achieving SDG6 suggests that the costs involved could be decreased by 25–40% through various efficiency measures. Smart water technologies and techniques have the potential to play a central role in attaining and even exceeding this.

References

ADB (2007) 2007 Benchmarking and Data Book of Water Utilities of India. Ministry of Urban Development, Delhi / Asian Development Bank, Manila.

Baby B and Vinod M S (2012) Research on self-sustained e toilet for households/urban-semi urban public/community sanitation. Presentation to FSM2, Faecal Sludge Management Conference, 29–31st October 2012, Durban, South Africa.

Carter R C and Ross I (2016) Beyond 'functionality' of handpump-supplied rural water services in developing countries. Waterlines, 35 (1), 95–110.

Colchester F E, Marais H G, Thomson P, Hope R and Clifton D A (2017) Accidental infrastructure for groundwater monitoring in Africa. Environmental Modelling and Software 91 (2017) 241–250.

Danilenko A, van den Berg, C, Macheve, B and Moffitt L J (2014) The IBNET Water Supply and Sanitation Blue Book 2014. World Bank, Washington, DC, USA.

Eram (2017) Awake to a clean India with eToilet. Eram Scientific, Thiruvavanthapuram, India.

Fisher M B, Mann, B H, Cronk R D, Shields K F, Klug T L and Ramaswamy R (2016) Evaluating Mobile Survey Tools (MSTs) for Field-Level Monitoring and Data Collection: Development of a Novel Evaluation Framework, and Application to MSTs for Rural Water and Sanitation Monitoring. International Journal of Environmental Research and Public Health 2016.

Gasson C (2017) A New Model for Water Access: A global blueprint for innovation. Global Water Leaders Group, Oxford, UK.

GLASS (2017) Financing Universal Water, Sanitation and Hygiene under the Sustainable Development Goals. UN-Water Global Analysis and Assessment of Sanitation and Drinking Water, GLASS 2017 Report, WHO, Geneva, Switzerland.

GSMA (2015) Mobile for Development Utilities Programme: The Role of Mobile in Improved Sanitation Access. GSM Association, London UK.

GSMA (2016a) The Mobile Economy India 2016. GSM Association, London UK.

GSMA (2016b) The Mobile Economy Asia Pacific 2016. GSM Association, London UK.

GSMA (2016c) The Mobile Economy Africa 2016. GSM Association, London UK.

GSMA (2017) State of the Industry Report on Mobile Money. GSMA, London, UK.

GWI (2016) Chart of the month: Utility vs discretionary spending. Global Water Intelligence, 17 (11), p. 5.

Hussain M (2015) 'We want to turn poo into gold': how SMS is transforming Senegal's sanitation. The Guardian, 12th August 2015.

Hutton G and Verguhese M (2016) The Costs of Meeting the 2030 Sustainable Development Goal Targets on Drinking Water, Sanitation, and Hygiene. World Bank, Washington DC, USA.

IFC (2009) Safe Water for All: Harnessing the Private Sector to Reach the Underserved. IFC, Washington DC, USA.

ITU and Cisco (2016) Harnessing the Internet of Things for Global Development.

JMP (2015) Progress on Sanitation and Drinking Water: 2015 Update and MDG Assessment, JMP UNICEF/WHO, Geneva, Switzerland.

JMP (2012) Progress on Drinking Water and Sanitation: 2012 Update, JMP UNICEF/ WHO, Geneva, Switzerland.

JMP (2008) Progress on Drinking Water and Sanitation: Special focus on sanitation, JMP UNICEF/WHO, Geneva, Switzerland.

Kolker J E, Kingdom W, Trémolet S, Winpenny J and Cardone R (2016) Financing Options for the 2030 Water Agenda. Water Global Practice Knowledge Brief. World Bank Group. Washington, DC, USA.

Krolikowski A, Fu X and Hope R (2013) Wireless Water: Improving Urban Water Provision Through Mobile Finance Innovations. University of Oxford, Oxford, UK.

Lloyd Owen D A (2016) InDepth: The Arup Water Yearbook 2015–16, Arup, London, UK.

McIntosh A C (2003) Asian Water Supplies: Reaching the Urban Poor, Asian Development Bank and IWA Publishing: London.

McIntosh A and Gebrechorkos S H (2014) Partnering for solutions: ICTs and Smart Water Management. ITU/UNESCO, Geneva, Switzerland.

Minh T (2016) Smartphone to detect water leakage. Vietnam News, 10th July, 2016.

Nique M and Smertnik H (2015) Mobile for Development Utilities Programme: The Role of Mobile in Improved Sanitation Access. GSMA, London, UK.

Ongole S D (2016) Presentation to Accelerating SMART Water, 6th SWAN Forum Annual Conference, London, 5–6th April 2016.

Oxford/RFL (2014) From Rights to Results in Rural water Services – Evidence from Kyuso, Kenya. Smith School of Enterprise and the Environment, Water Programme, Working Paper 1, Oxford University, Oxford, UK.

Prat M-A and Trémolet S (2013) An overview of sanitation app developments. Tremolet Consulting, London, UK Pump Aid (2013) 2012–13 Impact report. Pump Aid, London, UK.

Purvis K (2016) How do you solve a problem like a broken water pump? The Guardian, 22nd March 2016.

Rathbone D (2016) Mobile Coverage in the UK: Government plans to tackle 'mobile not-spots'. House of Commons Library briefing paper CBP-07069, HoC, London, UK.

Ross E (2016) Access to data could be vital in addressing the global water crisis. The Guardian, 27th October 2016.

RWSN (2009) Handpump Data 2009. Selected Countries in Sub-Saharan Africa, RWSN, St Gallen, Switzerland.

Sibthorpe C (2016). How a mobile can transform a woman's life. GSMA Connected Women. GSMA, London, UK.

Suez (2016) Suez wins a contract to improve water distribution services in a district of Kolkata, India. Press Release, 8th December 2016.

Thomas E and Matson K (2014) Monitoring with traditional public health evaluation methods: An application to a Water, Sanitation and Hygiene Program in Jakarta, Indonesia. Mercy Corps, Portland, USA.

Toilet Board Coalition (2016) The digitization of sanitation: Transformation to smart, scalable and aspirational sanitation for all. Toilet Board Coalition, Geneva, Switzerland.

TRAI (2004) Telecom Regulatory Authority of India, Press Release 1/2004.

TRAI (2008) Telecom Regulatory Authority of India, Press Release 11/2008.

TRAI (2011) Telecom Regulatory Authority of India, Press Release 11/2011.

TRAI (2017) Telecom Regulatory Authority of India, Press Release 12/2017.

van Beers P (2016) A Sustainable Business Approach is more effective and helps more people with the same amount of funding. FairWater Foundation, Amsterdam, The Netherlands.

Wesangula D (2016) The ATMs bringing cheap, safe water to Nairobi's slums. The Guardian, 16th February, 2016.

Winpenny J, Trémolet S, Cardone R, Kolker J E, Kingdom W and Mountford L (2016) Aid flows to the water sector: overview and recommendations. World Bank Group. Washington, DC, USA.

Winpenny J, Trémolet S and Cardone R (2016) Aid flows to the water sector: overview and recommendations. World Bank Group, Washington DC, USA.

WHO/UNICEF (2015) Estimates on the use of water sources and sanitation facilities. Updated June 2015, India. JMP, WHO/UNICEF, Geneva, Switzerland.

World Bank (2015) Leveraging Water Global Practice knowledge and lending: Improving services for the Nairobi water and sewerage utility to reach the urban poor in Kenya. World Bank Global Water Practice/WSP, Nairobi, Kenya.

Wyman O and JP Morgan (2016) Unlocking Economic Advantages with Blockchain: A Guide for Asset Managers. JP Morgan, New York, USA.

7

The Other 70%: Agriculture, Horticulture and Recreation

Introduction

Lower per capita water availability and increased demand for food are driving the need to improve the efficiency of agricultural and market garden irrigation systems and their application. Irrigation requires innovation to make the water that is available go further, to improve crop yields and to reallocate water to where it was not previously available.

7.1 Resource Competition and Municipal, Agricultural and Industrial Demand

Rising demand from municipal and industrial users is resulting in greater competition between water traditionally used for irrigation and these new consumers. Meanwhile, irrigation is facing a number of challenges. Firstly, only a finite amount of land is suitable for growing crops over a sustained period of time, and urbanisation is putting some of the most fertile of these lands out of productive use. Secondly, the growth in traditional agricultural yields is being overtaken by the growth in population and their consumption expectations. Finally, lowering groundwater levels and increasing soil salination is threatening the viability of irrigation agriculture in some areas.

7.1.1 Population Growth and Hunger Drive Demand

The conflict between users of water resources is a comparatively recent one, driven by shortages at the river basin level and in renewable groundwater resources. Irrigation accounted for roughly 70% of all water extracted in 2000 (FAO, 2010), having fallen from 89% in 1900 (Shiklomanov, 1999). Increasing abstraction for municipal, industrial and irrigation use can result in over-abstraction in areas where supplies were previously adequate. Areas where water has traditionally been scarce were often characterised by low population densities, adopting coping strategies such as transhumance [seasonal movement of people and livestock], or reliance on non-renewable groundwater resources.

Agriculture is set to be the greatest contributor to global water consumption in the medium to longer term (FAO, 2010). While municipal water consumption is forecast to rise from 600 billion m^3 pa in 2005 to 900 billion m^3 pa by 2030 and industrial demand

Smart Water Technologies and Techniques: Data Capture and Analysis for Sustainable Water Management, First Edition. David A. Lloyd Owen.
© 2018 John Wiley & Sons Ltd. Published 2018 by John Wiley & Sons Ltd.

from 800 billion m^3 pa to 1,500 billion m^3 pa, agricultural demand will rise from 3,100 billion m^3 pa to 4,500 billion m^3 pa. Global accessible renewable resources are assessed at 4,200 billion m^3 pa (700 billion m^3 pa for groundwater and 3,500 billion m^3 pa for surface water), a deficit of 2,800 billion m^3 pa net of those basins with a forecast surplus in 2030 (100 billion m^3 pa).

A rise in per capita agricultural water consumption from 194.5 m^3 per capita per annum in 1900 to 332.7 m^3 by 1960 was driven by higher standards of nutrition and more water-intense foodstuffs as well as the increased use of irrigation water. While there has been a fall in per capita consumption between 1960 and 2000 due to improved irrigation efficiency, population growth means that usage levels are still 52% higher than they were in 1900.

The greatest challenge lies in reconciling availability and reliability of supplies with an ever-increasing population and their new expectations, especially in developing economies where more water-intensive 'Western' diets are adopted. Water efficiency in agriculture rose by 1.0% per year between 1990 and 2004, compared with an average global population growth of 1.4% during this period (Winpenny et al., 2010). Demand for food is expected to rise by 38% from 2010 levels in 2030 (FAO, 2010) and 60% by 2050. Food consumption is also driven by the need to eliminate hunger. Globally, 850 million people were classified as undernourished in 2010 (FAO, 2011), or 14% of the population. In Pakistan and Bangladesh, the proportion being undernourished was 25% and 26% respectively and approximately 40% of people in South Asia are regarded as suffering from stunting.

Environmental conflicts are also emerging. In Europe, the EU water framework (2000/60/EC) and groundwater (2006/118/EC) directives are also impacting irrigation due to the need to preserve inland water flows and groundwater levels.

7.1.2 Loss of Productive Land

The amount of potentially useable land for agriculture is declining for a number of reasons. Unsustainable farming practices are impairing soil quality. For example, in river basins such as the Nile and the Indus, irrigated crop yields are declining as salt levels build up due to river flow modification preventing the occasional flooding required to wash excess salts away from the upper parts of the soil profile. Globally, there are 11.5 billion Ha of vegetated land; 1,660 million Ha (14%) being classified as lightly to moderately degraded, and 305 million Ha (3%) had suffered strong or extreme degradation by salination, topsoil loss or pollution, thereby being effectively beyond practical reclamation (Oldeman et al., 1991). In China, 4.5 billion tonnes of topsoil are lost every year due to erosion. Urbanisation is set to result in a projected cropland loss between 2000 and 2030 of 30 million hectares (range, 27–35 million) which is equivalent to 3.7% of the total crop production (range, 3.4–4.2%) due to the higher productivity of peri-urban crops (d'Armour et al., 2016).

7.1.3 Irrigation and Productivity

A total of 80% of water used for agriculture comes directly from rain, and about 20% comes from irrigation. Irrigated land accounts for 40% of the total crop yield. Yet irrigation is often wasteful, with 79% using the traditional flood (surface) method, 15% mechanised (spray) irrigation, and 6% sprinklers and drip (FAO, 2014b); these are broken

Table 7.1 Land equipped for irrigation by type in 2011.

Hectares (million)	
Equipped	324
Surface	280
Sprinkler	35
Localised (drip)	9

Sources: Adapted from FAO (2014a).

down in Table 7.1. In Asia, the proportion using irrigation is appreciably higher, mainly in South Asia and southern portions of China.

Drip irrigation is a somewhat recent development, with just 0.5 million hectares of what is also termed localised irrigation in 1981 (FAO, 2014a) (Table 7.2). As will be discussed later, it is the primary method when applying smart irrigation approaches.

The area equipped for irrigation has increased from 184 million hectares in 1970 and 258 million hectares in 1990, rising to 324 million hectares by 2012. Land may be equipped for irrigation, but the systems may not be used in a given year when there is enough rainfall not to require additional inputs. Actual irrigation took place on 275 million hectares in 2012, with 111 million hectares being irrigated by pumped groundwater. As more than one harvest can be taken in a year, the equivalent of 346 million hectares of crops was harvested from 261 million hectares of land in 2011 (FAO, 2014a, 2014b).

Asia (chiefly China and India) is the dominant region for irrigation (Table 7.2) both in terms of the amount of land irrigated as well as the intensity of irrigation in terms of its proportion of cultivated land, the crops harvested and its use of groundwater. In Europe and the Americas, rain-fed cultivation is the norm, with significant regional exceptions such as Southern Spain and California.

Irrigated land in developing economies is set to rise from 202 million Ha in 1999 to 242 million Ha by 2030 (FAO, 2010) and water withdrawal rising from 2,128 km^3 to 2,420 km^3, on the assumption that irrigation usage efficiency rising from 38% to 42%.

Table 7.2 Regional development of irrigation agriculture, 2011–12.

	% Of cultivated land irrigated	% Irrigated with groundwater	Irrigated crops harvested (million hectares)	Irrigated cropping intensity
Asia	41%	46%	271	141%
Americas	13%	39%	44	107%
Europe	9%	30%	15	100%
Oceania	7%	25%	2	100%
Africa	5%	18%	14	138%
World	21%	38%	346	130%

Sources: Adapted from FAO (2014a, 2014b).

Food exports (or 'virtual water' as the water consumed in one place is embedded in the exported product) also contribute to regional resource scarcity. Globally, groundwater depletion for agriculture rose from 194.7 billion m^3 pa in 2000, to 241.1 billion m^3 pa in 2010 with the groundwater depletion arising from exported foodstuffs rising from 17.7 billion m^3 pa in 2000 to 25.6 billion m^3 pa by 2010 (Dalin et al., 2017).

7.1.4 Irrigation Efficiency

Water is lost both when conveying it from the source to the field and in its delivery to the root growing area. Conveyance efficiency depends on the distance from the water source to the crops and the canal type. In 2010, 2,700 km^3 of water was abstracted to provide 1,500 km^3 of irrigation water, an efficiency of 56%. Irrigation efficiency is related to development; 48% in low income countries, 56% in middle-income countries and 61% in high income countries. Geography also plays a role, with an efficiency of 72% in Northern Africa against 26% in Sub-Saharan Africa (FAO, 2014a).

For unlined canals, the closer to the source, the more water gets conveyed to its delivery system. Likewise, the closer to the roots, the more effective is the delivery of water to the roots, as less is lost in both cases through evapotranspiration.

Poor maintenance can reduce the delivery efficiencies outlined in Tables 7.3 and 7.4 by up to 50%. In the very worst case, a poorly maintained sand canal of at least 2 km in length for surface (flood) irrigation could result in 7% of the water abstracted being productively used.

Table 7.3 Potential canal efficiency (%).

Conveyance type	Short (<2 km)	Long (>2 km)
Sand canal	80%	60%
Loam canal	85%	70%
Clay canal	90%	80%
Lined canal	95%	95%

Source: Adapted from Brouwer and Prins (1989).

Table 7.4 Potential application efficiency (%).

Application method	Efficiency
Surface	45–65%
Sprinkler	65–85%
Micro sprinkler	85–90%
Surface drip	85–95%
Subsurface drip	>95%

Source: Adapted from Irmak et al. (2011).

Table 7.5 Green spaces in Greater London.

Green spaces in London	% Of total land area
Public parks and gardens	5.8%
Playing fields	6.7%
Private gardens	14.0%
Total	26.5%

Adapted from GiGL (2010) and GiGL SINC (2015).

7.1.5 Urban and Domestic Irrigation

Urban areas have their own green spaces and these may generate a demand for irrigation, especially for parks, gardens and recreational areas. In some areas, garden irrigation and maintaining green spaces and playing fields are significant municipal water demand drivers. The irrigation these spaces need depends upon local circumstances, such as how they are used and rainfall patterns.

Data on urban green areas is inconsistent, even where it is gathered. The following are some illustrative examples. There are 432,964 hectares of gardens in the UK (Davies et al., 2009), along with 400,000 hectares of publicly accessible green space (HLF, 2016) and 12,000 hectares of King George playing fields, which were donated to municipalities in 1935–36 to mark the Silver Jubilee of King George V. For the Greater London area, more systematic data is available, as summarised in Table 7.5. London is regarded as one of the best provided major cities in terms of green spaces. Most cities will have considerably less green spaces.

Garden irrigation and municipal irrigation can account for a significant proportion of urban water consumption. Seven percent of domestic water consumption in England and Wales is used outdoors (Waterwise, 2012): 6% on garden irrigation and 1% for car washing. This is appreciably higher in the USA. Thirty percent of household water consumption in the USA is used outdoors: 16% on irrigation and the rest on car and path cleaning and swimming pools (US EPA, 2006), and 30% of overall municipal water use is for landscape irrigation (US EPA, 2006). Outdoor use is rising in some states, with the proportion in Texas increasing from 29% in 2004–08 to 33% in 2009–11 (Hermitte and Mace, 2012). A survey of 735 single family houses in California (DeOreo and Mayer, 2011) found that 53% of water consumption was for outdoor use, most being for plot irrigation. Twenty-five percent of household water in Australia was used outdoors in 2001, although this can be as high as 50% in Queensland (ABS, 2004). In Perth, 39% was for garden irrigation in 2008–09 (Water Corporation, 2010).

7.2 The Economics of Irrigation

In Chapters 3 and 4, the 'water-energy nexus' was considered in terms of the impact of water consumption on household energy bills and the potential to use these bills to influence customer behaviour. In a similar manner, the 'water and food nexus' is in part concerned with the value generated by water when used for irrigation when compared

Table 7.6 The impact of irrigation on water withdrawal and the economic contribution of agriculture in selected countries in 2013.

(2010 constant $)	Water withdrawal as a % of internal renewable resources	Irrigation as a % of total water withdrawal	Agriculture as a % of total GDP
Australia	5%	74%	2%
China	20%	65%	9%
Egypt	3,794%	86%	14%
India	53%	90%	15%
Saudi Arabia	986%	88%	2%
United States	17%	40%	1%

Source: Based on data extracted and adapted from the World Bank's WDI Database Archives (databank.worldbank.org).

Table 7.7 Value-added per unit of water for industry and agriculture compared in selected countries in 2013.

(2010 constant $)	Agriculture, value added ($ per m³)	Industry, value added ($ per m³)	Industry over agriculture value added per m³
Australia	1.58	131.69	83
China	1.83	27.99	15
Egypt	0.54	19.99	37
India	0.45	29.08	64
Saudi Arabia	0.64	511.35	805
United States	0.91	14.06	15

Source: Based on data extracted and adapted from the World Bank's WDI Database Archives (databank.worldbank.org).

with other applications. Tables 7.6 and 7.7 show water withdrawal for irrigation in its broader economic context in six countries. With the exception of the USA, irrigation is the dominant water use, while accounting for a small proportion of GDP. While it can be argued this reflects on how undervalued foodstuffs are, this lies outside the scope of this study.

The amount of value-added derived from each unit of irrigation water will be less than shown in Table 7.7, as it also includes value-added from non-irrigated crops and livestock. In Egypt, effectively all cultivated land value added comes from irrigated land (2004–2013 data used from FAO Aquastat). In the other countries, there is a mix of irrigation and rain-fed agriculture. For cultivated land (arable and permanent crops), the proportion being water managed areas varies from 6% (Australia), to 17% (USA), 44% (Saudi Arabia and India) to 60% (China).

Water allocations for agriculture are being reduced due to increasing scarcity. This is reflected by the rising cost of temporary and permanent water rights in the USA and Australia during periods of drought over the past two decades. In extreme cases, such as recently experienced in the Murray-Darling Basin in Australia, water allocations for agriculture are being completely withdrawn due to the effect of a multi-year drought. Murray Irrigation Limited regulates the provision of water to 2,400 farms in southern New South Wales. Spot entitlements (the right to use water over a pre-determined period of time) rose from an average of A$15.33 per megalitre (Ml) in 1998–99 to a peak of A$680.04 per Ml in 2007–08 before falling back to A$15.87 per Ml in 2011–12. Likewise, permanent entitlements rose from A$27–450 per Ml in 2008 to A$525–2,100 per Ml in 2008 and back to A$50–800 in 2012 (murrayirrigation.com.au). In the USA, water rights have evolved along a number of lines, including as an asset class in itself rather than being related to supplying a specific customer.

7.3   Smart Irrigation and Sustainability

Smart irrigation approaches address the inefficient watering of crops and amenity land by ensuring the greatest benefit is derived from the least water consumed. Water is only needed in those parts of the soil profile where roots are active and there is an evident need to avoid watering the soil when it is raining, or at times of day when it less effectively used. Irrigation regimes can therefore be realigned to optimising the soil moisture profile in relation to the ambient weather and root development.

Irrigation flow management is concerned with ensuring the optimal provision of water to plants both by controlling the flow of the water and by monitoring weather, soil and growing conditions to ensure no excess water is introduced to the crops. Examples such as AquaSpy, PlantCare and Dynamax for soil and weather monitoring systems will be considered. In addition, smart distribution can also be used to combine nutrient delivery with water delivery ('fertigation') to minimise the amount of fertilisation needed (and to lower the environmental impact of nutrient loading on the ambient environment) and to lower the cost of delivery through shared systems.

Other aspects include resource mobilisation and management for developing and delivering new water supplies where they are needed, including seawater mobilisation (DTI-r, dti-r.co.uk) for coastal (Seawater Greenhouse, seawatergreenhouse.com) and desert applications (Sundrop Farms, sundropfarms.com). Demand management is also being developed through lowering the actual amount of water needed by the plants by encouraging root growth in the vicinity of a limited but adequate availability of water. Both Eco-Ag (eco-ag.us) and DTI-r are adopting this approach. These processes are not 'smart' per se, but may provide platforms for smart water management systems.

Smart irrigation also uses fewer fertilisers and pesticides as well as less water, since their application can be synchronised. It also helps to avoid soil salination through minimising input and strategic soil flushing (White, 2013).

Many of the companies seen by the author in 2008–2012 are no longer active, suggesting that this is a market with a high attrition rate for new entrants; appreciably higher than that noted for smart domestic and municipal water and wastewater services.

7.3.1   The Market for Smart Irrigation

As with the market estimates and forecasts examined in Chapter 1.7.1, the market size estimates and forecasts below are best seen as means for considering how a market's development is being perceived by industry analysts. In market size terms, smart

Table 7.8 Global micro-irrigation market estimates and forecasts.

$ billion	Start year	End year	Start	End	CAGR
Research and markets (2017)	2015	2025	2.60	12.90	17.5%
Grand view research (2016)	2014	2022	2.48	8.76	17.1%
Mordor intelligence (2017)	2016	2022	3.11	8.07	17.2%

Sources: Adapted from Research and Markets (2017); Grand View Research (2016); and Mordor Intelligence (2017).

irrigation is an emerging sector within the irrigation market. The total global market for irrigation is estimated at $10–15 billion per annum in the mid-2000s (Alexander, 2008); new irrigation hardware $5 billion pa, and the replacement of older systems costing a further $5–10 billion pa. This does not include monitoring infrastructure.

There is a potential for drip irrigation to rise from its current level (9–10 million hectares) to 25% of the total irrigated area, or 80 million hectares, in the medium to longer term. Typical drip irrigation systems cost $1,000–3,000 per hectare, implying 70 million hectares of new drip irrigation systems at a capital cost of up to $70–210 billion (Wall, 2013). In India, drip irrigation systems cost RS 85,000 per hectare ($1,300 per hectare; Gangan, 2017). Some more specific irrigation market forecasts are summarised in Tables 7.8 and 7.9.

Smart irrigation uses micro-irrigation systems and drip feeders for the effective and controlled delivery of water. Micro-irrigation (Table 7.8) covers localised, programmable irrigation systems designed to deliver water to a specific location, including the control systems. This is mainly concerned with drip irrigation but also includes micro spray systems.

Drip irrigation (Table 7.9) covers the equipment used to apply controlled amounts of water either at the soil surface (in greenhouses, sometimes to the plant pots) or directly to the root zone. This does not include the monitoring and control systems. Drip irrigation is increasingly being used to efficiently deliver fertilisers at the same time. The increase in estimated market size and growth in the 2017 Markets and Markets survey over the previous survey suggests a recent uplift in demand (Markets and Markets, 2016a and 2017).

Table 7.9 Global drip irrigation market estimates and forecasts.

$ billion	Start year	End year	Start	End	CAGR
Markets and Markets (2016a)	2015	2020	2.14	3.56	10.7%
Research Nester (2017)	2016	2023	2.10	4.30	10.5%
Credence Research (2016)	2014	2022	1.07	2.75	12.5%
Markets and Markets (2017)	2017	2022	3.78	6.54	11.6%

Sources: Adapted from Markets and Markets (2016a, 2017); Research Nester (2017); and Credence Research (2016).

Table 7.10 Global smart irrigation market estimates and forecasts.

$ billion	Start year	End year	Start	End	CAGR
Markets and Markets (2016b)	2016	2022	0.50	1.50	17.2%
Statistics MRC (2017)	2015	2022	0.47	1.51	18.1%
Research and Markets (2016)	2015	2025	0.44	2.00	16.4%

Sources: Adapted from Markets and Markets (2016b); Statistics MRC (2017); and Research and Markets (2016).

Smart irrigation (Table 7.10) covers the hardware and software used to monitor soil moisture (or sap flow) and climate and to integrate this data with the other information needed to optimise the amount of water needed and to control the water delivery. Global sales of smart irrigation control systems were estimated at $100 million in 2011 (Aquaspy, 2013), along with $100 million for non-smart allied elements; $30 million for soil moisture monitors and $10 million for 'fertigation' (combining fertiliser and irrigation) and $70 million for greenhouse control systems.

Two sets of smart irrigation costs have been identified: firstly for accessing an external monitoring service or data handling service; and for the hardware, including the monitoring system and a data transmission system, and solar powered units where needed. The cost of the system is driven by the intensity of monitoring needed. A large area with a single crop and consistent soil, hydrographic and climatic conditions will need a lower monitoring density than more heterogeneous farmlands. Crop values and water and nutrient needs will also drive monitoring density. Examples cited by AquaSpy (Moeller, personal communication, 2010) show a range for system hardware costs of $140–1,450 per hectare (probes, weather monitors, data communications and terminals) with data services costing from $3.2–40.0 per hectare per annum. Annual fees for weather-based irrigation systems range from $48–360 (US DOI, 2012). The cost of a soil sensor (a probe and a controller) in 2009 was between $150 and $500 per unit (Cardenas and Dukes, 2016a).

The expression 'evapotranspiration' is widely used (Tables 7.11 and 7.12), but it is often misunderstood. Evaporation is non-productive loss, while transpiration is beneficial, as this covers water that has been used by the crop plants while growing. In many cases, estimates of 'evapotranspiration' only cover evaporation.

7.3.2 Policy Drivers

The role of policy as a driver for the adoption of smart water approaches is discussed in some detail in Chapter 8. Most policy impacts are the indirect result of water conservation or funding measures which may be effectively addressed through smart approaches. Policy itself is usually driven by external factors, principally through water scarcity, either in response to a period of drought or through the need to address underlying scarcity. In India, the irrigation technique has been the subject of a policy initiative, as the state of Maharashtra had mandated that at least 50% of its sugar cane cropland will

Table 7.11 Weather control unit costs.

Device	Application	Cost
Controller	Domestic (12 zones)	$275–1,200
Controller	Commercial (24 zones)	$1,195–2,800
Evapotranspiration gage		$1,375
Evapotranspiration gage	Controller interface	$435
Rain gage		$114–575
Rain gage	Controller interface	$435
Wind gage		$480–545
Wind gage	Controller interface	$435
Flow meter	One inch	$575
Smart control upgrade		$850

Adapted from US DOI (2012).

Table 7.12 Soil moisture unit costs.

Device	Application	Cost
Soil moisture sensor	Each	$180–290
Soil temperature sensor	Each	$98
Soil moisture controller	Domestic (12 zones)	$425–457
Soil moisture controller	Commercial (24 zones)	$1,097–4,080
Flow meter interface	One inch	$600

Adapted from US DOI (2012).

use drip irrigation by 2019 against 26% in 2017 (Gangan, 2017). In the USA individual states have made water reduction targets, which focus on irrigation, while in Australia, examples of irrigation caps have been identified. Where the cost of water provision has been increased beyond a certain point and this water is metered, this can act as an indirect driver towards smart irrigation, especially where consumers are able to appreciate the linkage between irrigation practices and water bills.

Where water rights are traded, more efficient irrigators will typically be less dependent on water rights, either buying fewer or none at all. They may also be able to sell on their water entitlements to third parties. The principal markets for traded water rights are in Australia (Murray-Darling basin) and the USA (Texas and the western states). Interest in trading water rights as an asset class, rather than for irrigation can distort these markets and how the water rights are valued.

Direct policy drivers to date has been by four states in the USA, which have for example mandated that smart pump controllers are to be used in garden and recreational land irrigation. In addition, grants for installing smart irrigation systems have been made in Texas. These are discussed in more detail in Chapter 8.

7.4   Smart Irrigation Agriculture

Agriculture may be perceived as a conservative business, yet when it needs to respond to external pressures, growers can and will do so. Sustenance agriculture excepted food is a commodity and growers compete to obtain the best prices for their products for the lowest cost of production. There are no direct incentives for using smart irrigation without price or regulatory pressure. Price pressure can be direct, when a grower has to pay for irrigation water or right to access water or indirect, when driven by the cost of pumping groundwater. There may also be restrictions on the amount of water that can be abstracted from a river at times of drought. Finally, the potential to deliver improved yields through optimised watering can also be a motivation in itself, especially when this also relates to better crop quality. For example, irrigation control can play a major role in the quality (and characteristics) of wines in a vineyard and indeed, getting the right balance between grape yield and quality. Finally, regulated deficit irrigation (Cooley et al., 2009) where less water is used than normal can be beneficial where the reduction in water use outweighs lower yields (Geerts and Raes, 2009) but this requires close monitoring to ensure plant health is not impacted.

7.4.1   Smart Irrigation Systems

At the simplest level, such as garden irrigation, this concerns a unit controlling irrigation water flow and timing with a basic link to climate data. At its most sophisticated (for crops and vineyards), this involves a detailed appreciation and analysis of growing needs and conditions, with irrigation flow managed in a number of separately watered zones. One challenge is the decrease in ground station weather monitoring globally since the 1990s due to lowered government support. Local initiatives are being developed to redress this, such as the 'Freestation' a fully automated weather station, developed with low cost components and free software that can be installed by the grower for $250. More complex weather stations are available, with examples noted by the US DOI (2012) costing up to $12,875.

As with municipal water management, smart irrigation consists of physical layers that are not smart per se and those that are involved in analysing and presenting the data obtained and responding to it. A smart irrigation system may receive data from a blend of internal sources and external sources where these have been subscribed to.

Smart irrigation systems are based on either monitoring soil moisture or rainfall. These can be combined and enhanced with further information about weather and soil conditions. In soil moisture systems, data is gathered from monitors that are distributed across the growing area, monitors measuring moisture levels in the vicinity of the root growing zone. The active root growth zone changes over time as the root system develops and moisture readings will need to reflect this, through a series of moisture monitors arranged down the probe. Some systems also have a deeper monitor to ensure that the soil is effectively flushed periodically to prevent soil salination. Readings are taken with capacitance probes (these can measure moisture, temperature and salination and other parameters, if needed) or tensiometers, which are used where higher moisture levels are needed. A variant on this is to record plant sap flow, as discussed in Case Study 7.1. In the weather-based systems, rainfall is recorded in terms of duration and intensity, along with other weather data as needed.

Soil and weather data is processed by a controller unit to determine when irrigation is needed and how much this needs to be modified for each zone. This data is married with information about the growing conditions, including the soil type, what crops are being grown, type of sowing (till or no-till) and degree of soil compaction. Data is usually presented both to track the parameters selected by the client and integrated into an irrigation schedule. The grower can either use the data to decide when to irrigate or it can be automated by remotely enabled irrigation timers. Weather forecasting is used to ensure that irrigation is avoided when rainfall is anticipated. Data analytics can include the time taken for water to infiltrate through the soil profile, comparisons of water usage over a set period, year to date and year by year, and irrigation performance, the amount of time when the actual soil moisture is at its optimal level, or too low or too high. Feedback loops based on this data enable the grower to refine the irrigation regimen to their local circumstances.

Water (and nutrients, where desired) is delivered to the growing zone through drip/tube irrigators. In some cases micro-sprays are used, but these are usually less efficient. Smart metering is employed in order to monitor and control the amount of water used. Thermal imaging can also be used to monitor the condition and effectiveness of the delivery system through pinpointing leaks. Each control unit will manage a number of irrigation zones, typically 4–12 for a domestic system and 8–48 for commercial applications. Some systems use a modular design to enable larger numbers of irrigation zones for particularly large or complex growing areas.

Irrigation also needs to be synchronised with each crop's growth cycle and the optimum times for irrigation in terms of a plant's needs and the length of the growth cycle (from planting to harvest). For example, sugarcane has a 365-day growth cycle and typically requires a total of 20,000 m^3 of water per hectare in 24 irrigations while maize has a 100-day growing season and requires 5,000 m^3 of water per hectare in six irrigations.

7.4.2 The Impact of Smart Irrigation

There have been two overviews of smart irrigation trials in the USA. A survey of irrigation management systems by the US Department of the Interior (US DOI, 2012) covering soil moisture sensors and rain sensors, and a more recent study by Williams, Fuchs and Whitehead (2014) which also included the impact of irrigation controllers on their own. Williams et al. (2014) reviewed 81 potentially applicable studies screening out reports which did not meet their criteria. The 34 trials that met their criteria were compared with irrigation control system. The US DOI report (US DOI, 2012) is from their *2012 survey of various commercial smart control systems which are retailed in the USA* (Table 7.13).

There was no significant difference noted between experimental and real-world trials, with savings of 13% and 16% for control systems, 39% and 37% for soil moisture monitors and 22% and 19% in experimental and real-world trials respectively (Williams et al., 2014).

Table 7.14 outlines some more recent trials by crop and location and also covers the potential for improved yields under optimal irrigation levels.

In California, watering costs were reduced by 75% from $47,336 to $11,834 per annum for 900 growing avocado trees through soil moisture monitoring allied with automated irrigation. 44 soil moisture monitors were installed at a cost of $8,200 into 22 irrigation

Table 7.13 Optimal water regimen irrigation, surveys.

System	WFU – savings	WFU – range	WFU – trials	US DOI
Irrigation controllers	15%	−35% to 43%	17	N/A
Soil moisture monitors	38%	4% to 72%	11	24–68%
Rain sensors	21%	13% to 34%	6	16–58%

Adapted from Williams, Fuchs, and Whitehead (2014) and US DOI (2012).
WFW refers to Williams, Fuchs and Whitehead (2014).

Table 7.14 Optimal water regimen irrigation, recent trials.

Location	Crop	Yield	Water saving	Source
Switzerland	Brussels sprouts	−11%	42%	PlantCare (2014)
Saudi Arabia	Tomatoes	4%	16%	Al-Ghobari (2014)
Saudi Arabia	Tomatoes	14%	26%	Mohammad et al. (2013)
USA	Cotton	31%	15%	Aquaspy (2008)

blocks. One sensor is placed 20 cm into the soil to measure moisture at the rooting level and a second 60 cm down to ensure enough water is used to prevent salt accumulation. Savings are anticipated to ease to 50% when the trees are mature, which is still cost-effective (Water Active, 2016).

As capital spending is need both for the irrigation as well as the monitoring systems, a financial incentive may be needed beyond purely improving the yield. Even so, where water is abstracted free in Spain for growing strawberries, a free smart water management app distributed by the beverages maker Innocent has gained acceptance as it lowered pumping costs. Here, water consumption data was gathered by Innocent (innocentdrinks.co.uk), a UK based beverages manufacturer from 2010 to 2012 to quantify each grower's water consumption and how consumption could be reduced while maintaining yield or quality. In 2014, the Irri Fresa app was launched, which alerts growers about optimal irrigation times. Innocent's participants reduced their water consumption by up to 40%, meaning that 1.7 million m^3 less water was used in 2015. During 2016, two other food brands and six retailers joined the Doñana Strawberry and Sustainable Water Management Group, with the aim of making water-efficient growing the norm.

Aquaspy a USA/Australian company (aquaspy.com) has developed a software system for the remote monitoring of soil moisture that allows for the optimal introduction of water and nutrients into the soil for a variety of plant growing applications. The company notes clients attaining 20–64% savings in their water use, with crop yields improving at the same time. Orange-co of Florida, USA pays $22,000 to use the system, which monitors soil moisture at 10-cm intervals down to 100 cm every 15 minutes. Savings of $300,000 per season came from lower water pumping costs and fertiliser applications. The Salmon Gum Estate vineyard in Australia doubled its wine yield and halved water usage using probes at five points down the soil profile.

Table 7.15 Regulated deficit irrigation regimen.

Location	Crop	Yield	Water saving	Source
Switzerland	Brussels sprouts	−9%	68%	PlantCare (2014)
USA	Wine grapes	3%	57%	Scholasch (2014)
USA [1]	Wine grapes	−25%	47%	Cooley et al. (2009)
USA [1]	Almonds	−4%	20%	Cooley et al. (2009)

7.4.3 Regulated Deficit Irrigation

A more drastic approach is to provide less water than would normally be considered as ideal. This approach, usually called regulated deficit irrigation (RDI; Table 7.15), works on the principle that crop quality can be improved when certain plants are water stressed, which has long been appreciated by oenologists [winemakers]. Because there can be a fine line between a beneficial deficit and harming a plant, close monitoring is needed and this favours species with a greater degree of drought tolerance such as vines, almonds and pistachios. Measuring sap flow rather than soil moisture, allows for a better understanding of the crop's actual health (Case Study 7.1).

The wine grape trial results reported by Scholasch (2014) are based on trials measuring sap flow in the vines (Case Study 7.1) while PlantCare (2012) and Cooley, Christian-Smith and Gleick (2009) surveyed trials using soil moisture monitoring.

7.5 Lawns, Parks and Sports Fields

Where municipal water supplies are used, there is usually a significant economic benefit from reducing irrigation water consumption. In the USA, a tradition of high water usage in drier areas means that the relatively low water tariffs are accepted. For example, domestic water use per capita per day in California was 469 litres, 519 litres in Texas and 530 litres in Arizona (Kenny et al., 2009). This compares with 125 litres in the Netherlands, 131 litres in Denmark and 150 litres in England and Wales in 2005–07 (Aquaterra, 2008). Since 2000, regional water shortages have prompted affected states in the USA to mandate the adoption of domestic irrigation water conservation devices including smart controllers in California and Texas. In Australia, domestic and recreational irrigation can be controlled by annual limits. These will be considered in more detail in Chapter 8. Where sports pitches and leisure facilities such as golf courses have their own water supplies, the main incentive for managing water consumption comes from reduced pumping bills.

Recreational and horticultural applications are distinct from irrigation agriculture. Domestic irrigation usually involves a heterogeneous landscape (trees, lawn and flower beds for example) within a relatively small plot, while sports and leisure irrigation are driven by how specific pitches and play areas are used and stand up to wear, along with considerations such as frequency of mowing. For these applications, plant yield is usually secondary to water consumption. Irrigation needs to reflect specific factors such as the resilience of the grasses used to water scarcity under different weather conditions.

Table 7.16 Playing fields and garden irrigation trials.

Location	Application	Water saving	Source
Florida, USA	Field – Potable	63%	Cardenas and Dukes (2016a)
Florida, USA	Field – Reclaimed	59%	Cardenas and Dukes (2016a)
Florida, USA	Residential – Reclaimed	44%	Cardenas and Dukes (2016b)
Colorado, USA	Residential – Potable	27%	Qualls et al. (2011)
Florida, USA	Residential – Potable	44%	Davis and Dukes (2015)

Another aspect of irrigation management is to ensure that playing surfaces do not feel wet underfoot, where drip irrigation works more effectively than traditional spray systems.

A sample of trials are summarised in Table 7.16. Reclaimed water can affect soil moisture monitoring and as a result, it has smaller savings than potable water, but its overall impact is greater as potable water is not needed in the first place.

The WeatherTrak (hydropoint.com) urban landscape irrigation software developed by HydroPoint Data Systems of the USA draws data from 40,000 weather stations in the USA to minimise, park, playing field and garden watering, lowering water use, nutrient run-off and erosion. In 24 case studies, savings in water use of 14–82% were noted. The system reduced water use by 39% and saved $108,000 in utility bills during 2009 when installed across 12 campuses in the USA.

Aquaspy (2008) cites a residential trial (urban and landscape) where soil moisture monitoring resulted in 46 irrigations rather than the 162 planned, with a reduction of water consumption from 48,000 m^3 per hectare to 13,000 m^3 per hectare, a saving of 73% and lowering the urban irrigation cost from $50,000 per hectare to $15,000 per hectare. This is where a utility's water was used. For a golf course, using pumped groundwater, a similar management system resulted in 11 irrigations instead of 58, with a reduction of water consumption from 13,820 m^3 per hectare to 5,180 m^3 per hectare, a 63% reduction, along with 73% less mowing and pumping costs falling from $296 per hectare to $78 per hectare.

Education and training are an important element for all types of irrigation management. This especially applies with high water users. In trials at residential properties in Orange County, Florida, heavy irrigation water users were identified and invited to take part in a smart irrigation management programme. While weather-based controllers reduced irrigation by 18% where these were used with an allied education programme, irrigation was 32% lower than before. Similarly, for soil moisture systems, irrigation fell by 30% in the control group and by 42% where combined with education (Dukes et al., 2016).

Weather-based irrigation controllers (WBICs) emerged in the 1990s with 20 manufacturers offering WBICs by 2014 and covered by the US EPA WaterSense scheme. Compliant units avoid more than 5% excess irrigation across all landscape zones and offer the full range of inputs necessary for the effective adjustment to local conditions, along with settings capable of being reprogrammed to reflect seasonal changes and self-diagnosis. In addition, when a water utility mandates exceptional reductions in water consumption, the controller has to be capable of responding to these. Temperature, solar radiation and radiation data are collected from sensors and used to calculate evapotranspiration in real time. Alternatively, weather data is obtained from an off-site source and transmitted to the controller (Western Policy Research, 2014). Adjustments are carried out using a remote device such as a smart phone or a computer terminal. Two-way communications, alerting users to the controller's operating status are becoming more commonplace. Field trials carried out between 2001 and 2011 found 7–24% reductions in outdoor water use and 7–10% reductions in total household use, or 140–220 litres per household per day. Further savings can be generated through systems which more closely track changes in seasonal weather and water needs.

7.6 Case Studies

7.6.1   Case Study 7.1: Wine Growing in the USA.
7.6.2   Case Study 7.2: Remote Sensing of Customer Water Consumption.
7.6.3   Case Study 7.3: Etwater – An Integrated Garden Management System.

7.6.1 Case Study 7.1: Wine Growing in the USA

Because of the high revenues that can be generated by vineyards and the need to closely control growing conditions, wine makers are willing to invest in innovative water management approaches at an earlier point than most other irrigators. Fruition Sciences, a Californian company (fruitionsciences.com) monitors vine performance through a bracelet that is attached to the vine for measuring sap flow. The vine bracelet was developed by Dynamax (dynamax.com) the smart water arm of Jain Irrigation.

The company carried out trials at six vineyards in Paso Robles, Napa and Healdsburg, California. In each case, adjacent plots were managed using traditional vineyard irrigation and a regimen driven by data from the vines. For the monitored blocks, sap flow sensors were attached to two vines 25 meters apart on the same row. The monitored vines were only irrigated when a pre-determined water deficit was reached. While the traditional blocks were irrigated 6–30 times during the growing season, the Fruition monitored blocks were irrigated 0–5 times.

Yields were 2.93 tonnes per acre for the Fruition blocks and 2.83 tonnes per acre for the traditional blocks. Quality was also seen as better in one case allowing more wine to be bottled under premium labels than previously. A total of 26 mm of water was used against 60 mm for the traditional bocks, a 57% saving. Actual savings were higher due to less irrigation later in the season as the health of the Fruition vines was noted.

Fruition Sciences has developed a range of vineyard wine monitoring and management services including water management. It serves 200 vineyards with over 1,000 vineyard blocks using the sap monitors. Larger deployments have seen water use reductions of 50% (Pahlmeyer) and 54% (Halter Ranch). In the former case, up to 75% of vines are not now being irrigated. A typical application, using 40 vine monitors will cost $4,000–5,000. The system is currently being developed for other high-value crops such as almonds (Giles, 2014).

Principal source: Scholasch (2014).

7.6.2 Case Study 7.2: Remote Sensing of Customer Water Consumption

OmniEarth was founded in 2014 and has developed a water image database for the state of California. In April 2015, the state announced that there would be a mandatory 25% cut in utility water consumption in the wake of the ongoing drought. One of OmniEath's main customers has been Inland Empire Water Utilities (ieua.org) a utility in serving 870,000 people in an area of 242 square miles to the east of Los Angeles.

OmniEarth (omniearth.net) uses IBM's Watson Visual Recognition (ibm.com) service for analysing aerial and satellite imagery to estimate the demand for water on a property-by-property level based on the land each property has and how it is used. IBM's Watson Visual Recognition allows for the presence of swimming pools, lawns and other pertinent features on 150,000 images of plots of land to be identified in 12 minutes. Satellite data is being used at a 2–5 meter resolution with GSD multispectral imagery (Fish et al., 2015).

The base subscription ($0.30–0.75 per parcel) analyses land cover and water budget results by each land parcel. The standard subscription ($0.05 per square meter in addition to the base subscription) analyses the relation between water meter reading data and the water budget and highlights heavy water users. The Watson recognition platform was customised to identify roofed area, pools, grass, shrubs and gravel along with irrigated and non-irrigated areas in individual plots and to calculate their surface area. OmniEarth generates visual coverage of the Water Resource Management product to overlay a Google Map.

OmniEarth is being used to identify heavy water use and to link this to swimming pools along with grass and shrubs where heavy irrigation was taking place. This allows water utilities to contact individual customers and to advise them about minimising the number of times their swimming pools are drained less and to consider along with advice about switching some grass or shrub areas to gravel or rocks or to consider using more drought tolerant plant varieties. In addition, customers are encouraged to check for leaks on their property and to explore more effective ways of irrigating their gardens. Actual water usage can be compared with ideal water consumption, along with benchmarking of water consumption with their peers and how improved efficiency impacts a customer's water bills. In some places, individuals may also be able to see the efficiency estimate for their property through connections with water-tracking apps like Dropcountr (Case Study 4.13). The service is being developed to identify changes in land and to quantity these so as to inform utilities about the impact of their water conservation programmes over a given period of time.

Currently OmniEarth is seeking to develop a system to identify inefficient irrigation regimens, where a customer is watering a lawn without taking the grass's growth cycle into consideration; smart irrigation systems usually only take into account sprinkler timing and soil moisture. From here, OmniEarth aims to develop an agricultural service for synchronising crop irrigation with each crop's growth and harvesting cycle. In April 2017, EagleView (eagleview.com) acquired OmniEarth with the aim of integrating its aerial imagery and data analytics with OmniEarth's systems. One of the areas EagleView seeks to develop is the ability remotely monitor irrigation agriculture efficiency.

Principal source: IBM (2016).

7.6.3   Case Study 7.3: ETwater – An Integrated Garden Irrigation Management System

The ETwater Unity system (etwater.com, short for evotranspiration water) is a weather-based irrigation management system developed for commercial gardens (housing associations and retail and office gardens), larger domestic gardens and parks and recreation facilities. The company is also looking at opportunities for serving golf courses and vineyards. It consists of a controller hub, for the automatic updating of the property's irrigation schedule in response to current and forecast weather conditions and at the client's location and cloud-based service that allows customers to modify the system to their preferences on a smart phone or tablet via a dedicated app. The ETwater Unity system is designed to be open to third party inputs through a product development kit. This is an open innovation platform, using Open Source software which also enables the system to interact with other digital platforms.

The ETwater Unity app can be downloaded for free and is compatible with Apple and Android based mobile platforms. The app provides information on the user's irrigation water consumption and outlines the potential for reducing this from an ideal irrigation baseline that is driven by local conditions and user experiences. The app serves as a gateway to ETwater's paid for systems and services.

The ETwater Smartest Sprinkler Service costs from $35 per month. This provides an interactive, satellite-based image of the customer's garden on a smart device. Details about plants and irrigation systems are then added to the site, along with other pertinent information such as shaded areas. Once integrated with the controller and the irrigation system in the garden, watering is adjusted to meet the needs of the differing parts of the garden according to the plants present and the current and forecast weather conditions. Data is provided to show how much water has been used (and saved) over set periods of time and put into a meaningful context through real world examples for comparison. Water savings can also be compared with other subscribers in the area. A 'Municipal Restrictions' feature enables users to factor in any periods (time and day) when irrigation is being prohibited in their area into their irrigation schedules. The service is being developed to provide an image- recognition capability whereby customers take a photograph of their lawn and other grassland areas in order to obtain an analysis of the soil conditions.

Principal source: ETwater website.

Conclusions

From the trials carried out to date, it is evident that smart approaches towards avoiding excess irrigation can achieve significant savings. The generation of data on the effectiveness of smart irrigation remains at a somewhat early stage as trials to date have been carried out on a comparatively limited number of crops in a number of locations under various conditions. These trials suggest that savings of 21–34% for rain sensor based systems (a range of 13–58%) are attainable, along with 38–46% for soil moisture based systems (a range of 4–72%).

Water consumption is not the only driver. Yields can benefit from the prevention of over-watering and in some of crops (especially grapes grown for wine making) managing irrigation plays a significant role in optimising crop quality. Soil monitoring can prevent saline build-up by alerting when saline levels are too high so that salts can be flushed below the growth zone.

Irrigation control has been taken further by seeking to underwater a crop (regulated deficit irrigation, RDI) for more drastic savings, along with improved crop quality. Growing crops under stressed conditions requires more intensive monitoring in order not to damage the crops, and certain crops (grapes, almonds and pistachios for example) are appreciably more resilient in this regard. One approach for ensuring the effective use of RDI is to monitor the sap flow through a plant in order to appreciate the actual impact of a particular irrigation programme. The higher costs associated with sap flow monitoring mean that it is best suited to high-value crops, especially vines for premium wines.

To be effective, smart irrigation needs to be linked with efficient and targeted methods of water delivery, preferably drip irrigation. Currently, 3% of crop land equipped for irrigation uses drip irrigation, but the proportion has been increasing in recent years.

References

ABS (2004) Water Account Australia, 2000–01, Australian Bureau of Statistics, Canberra, Australia.

Aquaterra (2008) International comparison of domestic per capital consumption. Aquaterra, for the Environment Agency, Bristol, UK.

ADB (2013) Thinking about Water Differently: Managing the Water–Food–Energy Nexus. Asian Development Bank, Metro Manila, Philippines.

Alexander L (2008) Water Q&A: Secular Trends, Multiple Opportunities. Jefferies and Co., New York, NJ, USA.

Al-Ghobari H M (2014) The assessment of automatic irrigation scheduling techniques on tomato yield and water productivity under a subsurface drip irrigation system in a hyper arid region. WIT Transactions on Ecology and The Environment (185): 55–66.

d'Amour C B, Reitsma F, Baiocchi G, et al. (2016) Future urban land expansion and implications for global croplands. PNAS 2016: 1606036114v1-201606036.

Aquaspy (2008) Company overview. Aquaspy, Santa Ana, CA, USA.

Aquaspy (2013) Presentation to the World Water Tech Investment Summit, London 6–7 March 2013.

Beddington J (2013) The Perfect Storm; what is happening to the World? Lecture, 12th November 2013, Imperial College, London, UK.

Brouwer C and Prins K (1989) Irrigation Water Management: Irrigation Scheduling. Training Manual no. 4. Food and Agriculture Organization of the United Nations, Rome, Italy.

Cardenas B and Dukes M D (2016a) Soil moisture sensor irrigation controllers and reclaimed water part I: Field-plot study. Applied Engineering and Agriculture 32(2): 217–224.

Cardenas B and Dukes M D (2016b) Soil moisture sensor irrigation controllers and reclaimed water part II: Residential evaluation. Applied Engineering and Agriculture 32(2): 225–234.

Cooley H, Christian-Smith J and Gleick P (2009) Sustaining Californian Agriculture in an Uncertain Future. Pacific Institute, Oakland, CA, USA.

Credence Research (2016) Drip irrigation systems market by application (agriculture, gardens, others), component (drippers, tubing, backflow preventers, valves, pressure regulators, filters, fittings) – Growth, share, opportunities and competitive analysis, 2015–2022.

Dalin C, Wada Y, Kastner T and Puma M J (2017) Groundwater embedded in international trade. Nature, 30th March 2017 (545) 700–704.

Davies Z G, Fuller, R A, Loram A, Irvine K N, Sims V and Gaston K J (2009) A national scale inventory of resource provision for biodiversity within domestic gardens. Biological Conservation 142 (4): 761–771.

Davis S and Dukes M (2015) Methodologies for successful implementation of smart irrigation controllers. Journal of Irrigation and Drainage Engineering 141(3): 04014055.

DeOreo W B and Mayer P W (2011) California Single Family Water Use Efficiency Study. California Department of Water Resources, Aquacraft, Boulder, CO, USA.

Dukes M D, Allen L M, Thill, T, et al. (2016) Smart Irrigation Controller Demonstration and Evaluation in Orange County. Water Research Foundation, Denver, CO, USA.

Fish C, Slagowski S, Dyrud L, et al. (2015) Pull vs. Push: How OmniEarth Delivers Better Earth Observation Information to Subscribers. The International Archives of the Photogrammetry, Remote Sensing and Spatial Information Sciences Volume XL-7/W3, 2015. 36th International Symposium on Remote Sensing of Environment, 11–15 May 2015, Berlin, Germany.

FAO (2010) Towards 2030/2050. UN FAO, Rome, Italy.

FAO (2011) The State of Food Insecurity in the World 2011: A Report by the High Level Panel of Experts on Food Security and Nutrition of the Committee on World Food Security. UN FAO, Rome, Italy.

FAO (2014a) Area equipped for irrigation. Aquastat, Food and Agriculture Organization of the United Nations, Rome, Italy.

FAO (2014b) Irrigated crops. Aquastat, Food and Agriculture Organization of the United Nations, Rome, Italy.

Gangan S P (2017) To save water, Maha government wants 50% sugarcane crop on drip irrigation in 2 years. Hindustan Times, 15th May 2017.

Geerts S and Raes D (2009) Deficit irrigation as an on-farm strategy to maximize crop water productivity in dry areas. Agricultural Water Management 96: 1275–1284.

Giles E (2014) Grow more food with less water? There's an app for that. The Guardian, 25th September 2014.

Greenspace Information for Greater London (2015) GiGL, London, UK.

GiGl (2010) London: Garden City? Greenspace Information for Greater London, London Wildlife Trust and Greater London Authority, London, UK.

Grand View Research (2016) Micro-Irrigation Systems Market Analysis by Product (sprinkler, drip, center pivot, lateral move), by crop (plantation crops, orchard crops, field crops, forage and turf grasses) and segment forecasts to 2022.

Hermitte S and Mace R E (2012) The Grass Is Always Greener...Outdoor Residential Use in Texas. Texas Water Development Board, Austin, TX, USA.

HLF (2016) State of UK Public Parks 2016. Heritage Lottery Fund, London, UK.

IBM (2016) OmniEarth Inc: Combating drought with IBM Watson cognitive capabilities. IBM, New York, USA.

Irmak S, Odhiambo L O, Kranz W L and Eisenhauer D E (2011) Irrigation Efficiency and Uniformity, and Crop Water Use Efficiency. Biological Systems Engineering: Papers and Publication, University of Nebraska.

Kenny J F Barber N L, Hutson, S S, et al. (2009) Estimated Use of Water in the United States in 2005. US Geological Survey, US Department of the Interior, Reston, VA, USA.

McKinsey (2009) Charting Our Water Future: Economic frameworks to inform decision-making. World Bank 2030 Water Resources Group, Washington DC, USA/McKinsey, London, UK.

Markets and Markets (2016a) Drip Irrigation Market by Crop Type (field crops, vegetables, orchard crops, vineyard and rest (crops)), application (agriculture, landscape, greenhouse and others), components and by region – Global trends and forecasts to 2020.

Markets and Markets (2016b) Smart Irrigation Market by Irrigation Controller (weather-based controllers and sensor-based controllers), hardware and network components, application, and geography – Global forecast to 2022.

Markets and Markets (2017) Drip Irrigation Market by Component (emitters, drip tubes/drip lines, filters, valves and pressure pumps), emitter/dripper type (inline and online), application (surface and subsurface) crop type and region – Global forecast to 2022.

Mohammad F S, Al-Ghobari H S and El Marazky M S A (2013) Adoption of an intelligent irrigation scheduling technique and its effect on water use efficiency for tomato crops in arid regions. Australian Journal of Crop Science 7(3): 305–313.

Mordor Intelligence (2017) Micro-Irrigation Systems Market – Global Trends, Industry Insights and Forecasts (2017–2022).

Oldeman L R, et al. (1991) World Map of the Status of Human-induced Soil Degradation (GLASOD). UNEP, Nairobi, and ISRIC, Wageningen, The Netherlands.

PlantCare (2014) An impressive result. PlantCare, Russikon, Switzerland.

Qualls R J, Scott J M and DeOreo W B (2001) Soil moisture sensors for urban landscape irrigation: Effectiveness and reliability. Journal of the American Water Resources Association 37(3): 547–559.

Research and Markets (2017) Global Microirrigation Systems Market Analysis and Trends 2013–2016 – Industry Forecast to 2025.

Research and Markets (2016) Global Smart Irrigation Market Analysis and Trends – Industry Forecast to 2025.

Research Nester (2017) Drip Irrigation Market: Global Demand, Growth Analysis and Opportunity Outlook 2023.

Shiklomanov I A (1999) World water resources and their use. UNESCO, Geneva, Switzerland.

Scholasch T (2014) A comparative study of traditional vs plant-based irrigation across multiple sites: Consequences on water savings and vineyard economics. Application during a drought in California. Fruition Sciences, Oakland, CA, USA.

Statistics MRC (2017) Global Smart Irrigation Market 2017: Share, Trend, Segmentation and Forecast to 2022.

US DOI (2012) Weather and Soil Moisture-Based Landscape Irrigation Scheduling Devices. Technical Review Report, 4th Edition. US Department of the Interior, Southern California Area Office, Temecula, CA, USA.

US EPA (2006) Outdoor Water Use in the United States. EPA WaterSense, US EPA, Washington DC, USA.

Wall M (2013) Smart Water: Tech guarding our most precious resource. BBC, 30th July, 2013.

Water Corporation (2010) Perth residential water use study 2008–2009, Water Corporation, Perth, WA, Australia.

Waterwise (2012) Water – The Facts. Waterwise, London, UK.

WEF (2011) Water Security: The Water–Food–Energy–Climate Nexus. World Economic Forum Water Initiative, Geneva, Switzerland.

Western Policy Research (2014) BMP Cost and Savings Study Update: A Guide to Data and Methods for Cost-Effectiveness Analysis of Urban Water Conservation Best Management Practices, July 2014 Update. Western Policy Research, Santa Monica, California, USA.

Williams A, Fuchs H and Whitehead C D (2014) Estimates of Savings Achievable from Irrigation Controller. Lawrence Berkeley National Laboratory, Berkeley, CA, USA.

Winpenny J, et al. (2010) The wealth of waste: The economics of wastewater use in agriculture. FAO water reports 35, UN FAO, Rome, Italy.

8

Policies and Practicalities for Enabling Smart Water

Introduction

Policy involves the development of principles or rules which are intended to achieve rational outcomes. In terms of public policy this means the development of rules, laws or directives affecting aspects of public life. Policy can be implemented either as guidelines which are meant to encourage the achievement of an intended outcome or as mandatory instruments which aim to compel the intended outcome to occur.

Policy and smart water are rarely mixed by choice. It will become evident that the practical role of policy in smart water development is to assist in creating the conditions needed for its development and both direct and indirect adoption. That also reflects smart water's role as a tool for practitioners when addressing pre-existing cost, supply and management challenges.

8.1 Regulation as a Policy Driver

Public health and environmental obligations, driven by laws, directives, standards and regulations motivate water utility mangers to monitor a variety of chemical and physical parameters through the water distribution network and to ensure that they comply with the standards in place. There is an increasing drive towards rapid or ideally realtime data capture and analysis in order to prevent or minimise the impact of any perturbations and to provide access to this data. In addition, a new generation of high speed bacterial testing systems are enabling some biological and biochemical criteria to be monitored effectively in realtime.

Regulations such as the EU Water Framework Directive (WFD, 2000/60/EC) are encouraging a shift from reactive to proactive inland water quality management, which is creating demand for realtime collection and assimilation of physical, chemical and biochemical data. The need to reconcile maintaining river water flow under the WFD with customer needs in turn drives demand management and smart metering, as is the requirement for water utilities to be charged full cost recovery tariffs (in theory at least, since 2010), which encourages customers to consider their water usage. Applications include laboratory sensors for drinking water quality, governed by World Health Organization standards (set out in its 'Guidelines for drinking-water quality', 4th edition,

Smart Water Technologies and Techniques: Data Capture and Analysis for Sustainable Water Management, First Edition. David A. Lloyd Owen.

2011) and national standards, along with the Drinking Water Directive 98/83/EC in the EU; on-site testing for bathing water quality and rapid dissemination of test results for the EU's Revised Bathing Water Directive (2006/7/EC); and the analysis of discharges from wastewater treatment works and industrial sites as regulated by the Urban Wastewater Treatment Directive (97/271/EEC); and the Integrated Pollution Prevention and Control Directive (2008/1/EC) in the EU.

When it comes raising funds and project development, companies, universities, and funders need some evidence that markets do or will exist to justify developing smart water systems. Direct policy interventions include cases where governments have specified that a smart water approach should be adopted, such as smart water meters. Indirect policy incentives include tariff policies that encourage demand management along with water and wastewater quality and service delivery standards that are most effectively met through realtime monitoring and management.

8.2   Direct Policy Interventions

The direct or explicit encouragement of the development and deployment of smart water systems has been seen in Malta, Australia, Korea, the USA (at the State level), the United Kingdom (in Jersey) and Singapore. In Canada (Ontario) and Israel, policy has been directed in supporting smart water technology companies.

In Jersey and Malta this has seen the roll-out of a universal smart water metering programme. Both Malta and Jersey adopted smart water metering in order to respond to water shortages without developing new desalination plants. In Malta, smart water metering was adopted when it was realised that it would be more cost-effective to install these at the same time as smart electric meters which are to be deployed by 2020 in response to EU legislation (2009/72/EC). In Australia, the government funded a community-wide smart water metering trial. Korea and Singapore have sought to develop comprehensive smart water grids.

8.3   Indirect Policy Interventions

Policies may have unintended consequences. In Scotland, competition for non-domestic customers was introduced in 2008 and triggered the deployment of smart meters, as utilities and utility retailers sought to differentiate themselves from their peers through improved performance and service delivery.

Smart water meter use can result from measures designed to encourage efficiency. For example, policies designed to lower water usage in California and Arizona have seen utilities adopting smart water meters to inform customers about their water usage. In Denmark, a water supply tax introduced in 1994 charges water utilities €1 per m^3 for any water above a 10% threshold that is lost to leakage (Fisher, 2016). This subsequently drove the adoption of smart leakage detection systems in Copenhagen. In Australia, the annual irrigation allowance in Canberra for parkland, sports grounds and residential gardens is 0.5 Ml per 1,000 m^2 of total surface area per annum or 5,000 m^3 per hectare per annum (ACT, 2007). The limit-based nature of this legislation is driving both the adoption of smart metering and smart irrigation.

The adoption of performance measures that accurately reflect the status of a system is an important element towards enabling their quantitative analysis. One of the recurring themes in this study has been the assumed necessity for using percentage figures for water loss when more rational measures are available. This reflects the tension between reporting meaningful information and expressions that are popularly adopted. The infrastructure leakage index (ILI), the ratio of actual real losses over the minimum real losses in a system aims to reflect how the efficiency of a network changes irrespective of the water flowing through it, which cannot be done with percentage loss numbers. ILI is starting to gain international acceptance. ILI has been accepted by the main professional organisations worldwide since 2005, and was adopted by Malta and Austria before 2005 and Denmark, Croatia, Italy, Germany and Korea between 2008 and 2017 (Merks et al., 2017).

8.4 Policy as an Inhibitor

Until 2010, a lack of policy coherence in England and Wales regarding the pace and nature of water metering have inhibited smart metering and has effectively prevented the synchronisation of smart water and electrical meter programmes. This was exacerbated by a lack of incentives for innovation, as five-year spending cycles are seen as poorly suited for longer-term projects. Recent reflexions on water policies and the benefit of smart metering are belatedly changing this (Ofwat, 2012) as seen in the adoption of smart metering by utilities such as Southern Water and Thames Water (Chapter 4).

8.5 Policy Challenges

International cooperation on agreeing and adopting common standards for hardware and software, especially regarding their interoperability are needed to encourage the international deployment of smart water technologies. This is becoming increasingly important as various smart water applications, currently operated independently, start to be integrated (ITU, 2010). Smart metering standards are under consideration at the European level, including the EN 13757 standard for data transmission between a smart meter and a data concentrator. International standards for water data-communications and geographic information systems are being developed by the ISO (including ISO 22158 for water meter electronic interfaces, the ISO 27000 series for data security, ISO 37120 for smart cities and the ISO 55000 series for asset management) along with the ISO 19100 series for geographical information systems. The Danish Standards organisation is developing smart city standards in conjunction with the ISO and the International Electro-Technical Commission (Freedman and Dietz, 2017). The only national initiative for smart water standards identified to date is the Open IoT Standards set of guidelines developed by the Government of Singapore (Freedman and Dietz, 2017).

8.6 Case Studies

The seven case studies here are snapshots of how policy has (or has not) led to the development or deployment of smart systems and services. Six are examples of direct and indirect support. The UK was until 2017, distinctly ambiguous. A recent position paper by the economic regulator for England and Wales indicates that this is starting to change.

8.6.1 Case Study 8.1: Australia – Localised Initiatives

The 2003–10 drought focused attention on Australia's water demand and supply management priorities. The Australian Government's Water for the Future initiative was a ten-year programme managed by the Department of Sustainability, Environment, Water, Population and Communities to secure water supplies and maintain river water quality. It is based on the Water Act 2007 and subsequent amendments. 'Smart' water is not mentioned in the initiative. Even so, smart water was supported by two specific initiatives.

Government approval for smart systems: The Australian Government provided A$1.8 million to support a Smart Approved WaterMark Scheme (SAWM, see Chapter 4.5). The Smart Approved WaterMark is a label for water efficient outdoor products and services in order to assist consumers to choose water efficient goods. These included rain sensors provided by two companies and soil moisture sensors provided by three companies along with consultancies for agriculture and low water plant selections for gardens.

Supporting a smart metering roll-out: Wide Bay Water Corporation in Queensland installed an initial roll-out of 2,359 smart domestic and commercial water meters in 2006–07, with 24,500 meters covering over 60,000 people installed by 2010. The A$5.7 million project was supported by A$2.6 million from the national government and A$0.9 million from the state government. This enabled field trials to be carried out at the first substantial community in Australia that was wholly covered by smart meters. It is being used as a trial for Australian smart water meter applications (Turner, 2010) including leak alerts for customers (Freedman and Dietz, 2017).

8.6.2 Case Study 8.2: Ontario, Canada – A Smart Grid for Water

In Ontario, Canada, the Water Opportunities and Water Conservation Act of 2010 and the Ontario Water Technology Acceleration Program and Efficiency Initiative of 2012 seek to encourage customers to use water more efficiently by requiring standardised information about water use on water bills. The goal is to educate consumers on their water use, with enhanced customer data.

The Ontario Water Opportunities and Conservation Act, aims to coordinate regulation and policy with sustainable water management. This includes water sustainability plans, standardised information about water use on bills, water efficiency standards and a Water Technology Acceleration Project (WaterTAP) to support innovation by companies. The province supports VC funding the Ontario Venture Capital Fund, the Ontario Emerging Technologies Fund, the Investment Accelerator Fund and the Innovation Demonstration Fund and research through 150 researchers at 25 water-related research institutes. One example is the Ontario MaRS incubator, a public private partnership designed to increase the state's global competitiveness. It supported the development of Echologics (echologics.com) an acoustic-based large pipe leak-detection company founded in 2003 which was acquired by the Mueller Company (muellercompany.com) in 2011.

8.6.3   Case Study 8.3: Israel – Supporting Smart Technologies

Israel is a water-scarce country which has in recent years experienced droughts along with other potential threats to its water supplies. Water scarcity has created markets for water-saving technologies for domestic and municipal uses. Economic incentives designed to reduce water demand in the urban and agricultural sectors, based on increasing block tariffs resulted in the development of innovative water management devices, such as water meters that are read remotely and more accurately (in order to highlight leakages), pressure optimisers, and computerised irrigation systems (OECD, 2010b).

Developing smart technologies is supported by an inter-ministerial committee. Israel's Mekorot (mekorot.co.il) established the Water Technologies Entrepreneurship Centre (WaTech) in 2004 as a platform for business ventures. In its first three years, 250 projects were assessed and 35 of these were subsequently trialled by Mekorot. The utility sets technical standards at the start of the trial and their deployment is seen as a tightly structured process in order to minimise the potential for innovation risks.

In 2006, Israel launched the Novel Efficiency Water Technologies (NEWTech, israelnewtech.gov.il) programme which has seen in 26 government-funded water technology incubators, gaining $700 million in private investment. In 2008 this was rebranded the National Sustainable Energy and Water Program. NEWTech will reimburse 70% of installation costs (up to $200,000) in order to mitigate the risk in installing innovative technologies. Companies such as TaKaDu (leakage detection and management, takadu.com) and Miya (pressure management, miya-water.com) have emerged from this policy support. In 2017, the programme was involved with 151 companies involved with water systems and services.

8.6.4   Case Study 8.4: Korea – Smart Water as Part of a National Competitiveness Package

Korea aims to be one of the world's leading exponents of the smart water grid by 2020 through various R&D projects to develop a '3S' (Security, Safety, Solution) platform technology. Korea is seeking to link all aspects of water treatment and management including agricultural use and dams with data flows that mimic the water cycle, reporting to a centralised control facility. Two levels of smart grids are envisaged, a 'micro-grid' at the town level and a 'macro-grid' at the regional or river basin level. The PRIME technical strategy is designed to support the deployment of smart water systems: Platforms (water grid and information technology); Resources (secure and activate various water resources); Intelligent network (self diagnostic sensors and ICT backed co-operation network); Management (resource risk management and asset management); and Energy efficiency (smart water-energy grid) Choi and Kim (2011).

The policy support has come from the Science and Technology Basic Law. Article 13 includes a series of National S&T Basic Plans, the second running from 2008–12 and the third running from 2013–17. Science and Technology policy is administered by KISTEP, a government-funded science and technology policy planning and evaluation institution under the National Science and Technology Commission, which in turn reports directly to the President of the Republic. By 2015, approximately 100 parameters were being evaluated by the smart network trials (Choi et al., 2016).

The smart grid plans stem from a central plan to sustain economic development through international leadership in selected information technology related themes. Here the smart grid concept was seen as meeting both the country's water management needs and as a platform for gaining international business. The smart grid has been deployed on a trial serving Yeongjongdo Island (site of Korea's international airport) looking at how it can be used to manage potential water supply risks (Byeon et al., 2015). In Songdo a newly built city, smart water, energy, communications and transport networks were implemented by the Government from the onset, including comprehensive water and wastewater asset and event monitoring (Freedman and Dietz, 2017).

8.6.5   Case Study 8.5: Singapore – Smart Management as a Part of Holistic Water Management

Because of the City State's dependence on water imported from Malaysia and its aim for water self-sufficiency by 2060, water management is regarded as a priority by the Government. The state-owned Public Utilities Board (PUB) is responsible for all aspects of water management and service provision, under the ministry of Environment and Water. PUB is mandated to work with companies and research institutions for developing new technologies that can assist in water management and create opportunities for Singapore-based enterprises.

Using the utility as a smart technology testing platform: The PUB acts as a test bed for new technologies and their application via public private partnerships with the Economic Development Board being a facilitator, with the intellectual property remaining in the private sector. Amongst the 294 R&D projects carried out between 2002 and 2010, smart water projects have looked at realtime water quality monitoring and analysis, membrane integrity sensors for wastewater recovery and microbial source tracking (Cleantech Group, 2011). In 2015, a smart showerhead providing realtime consumption data was launched by Rigel Technology (rigel.com.sg) after being supported by the PUB (Public Utilities Board Singapore, 2016).

A smart grid to optimise water management: In order to attain water self-sufficiency in an affordable manner, all aspects of the water cycle are managed together including desalination, wastewater recovery, catchment management and rainwater sewerage recovery. This also requires extensive monitoring which is integrated into a smart water grid. Policy support has come from the 2009 Active, Beautiful, Clean Waters Programme (inland water quality) and a series of water efficiency initiatives including the 2011 Water Conservation Awareness Programme. Due to the efficiency of the extent water network (distribution losses are less than 5%) and the other monitoring programmes, household smart water metering has not to date been seen as a priority (Lloyd Owen, 2012). Smart metering (AMR) is currently being trialled at 1,200 properties, with the aim of selecting a preferred technology by 2018 and full coverage by 2028 (Public Utilities Board Singapore, 2016). Home water management systems (consumption logging and graphical data display) have been trialled at 3,200 properties since 2015 (Public Utilities Board Singapore, 2016).

Policy development in Singapore is aimed towards long-term water self-sufficiency. Adoption of smart water measures is pragmatic, being focused towards areas where it is seen to meet their needs. Where this is the case, policy is supportive especially in assisting the development of appropriate approaches.

8.6.6   Case Study 8.6: The United Kingdom – Mixed Signals

England and Wales – regulatory neutrality: Water and wastewater services in England and Wales were privatised in 1989. Ofwat, the sector's economic regulator has statutory powers to set price and performance limits, which in effect determine each company's spending priorities. The sector operates within a series of five-year Asset Management Plans (AMPs); AMP6 runs from 2015 to 2020 and AMP7 will run from 2020 to 2025.

Until 2010, Ofwat rejected attempts by utilities to install smart water meters, regarding this issue as too strongly linked with water scarcity and only allowed intensified manual meter roll-outs in the south east of England. 'Playing our part' Ofwat's climate change focus report (Ofwat, 2010) states that, 'an innovation platform for water…could mean extra funding' recognising that smart meters have a role in leakage detection and that leakage needs to be seen as an integrated element of supply–demand balance rather than a regulatory target (Worsfold, 2012). Future monitoring may include flexible or seasonal targets, implying a potential need for smart metering. Since 2011, Defra has acknowledged the importance and benefits of smart water metering, but considers that it is premature, economically and financially, to commit to such schemes (MacDonald, 2010). Even when meters are installed by a utility, volumetric tariffs can only be charged when the customer opts for them.

In 2011, Ofwat concluded that the greatest net benefit would be delivered from an accelerated programme moving from 38% metering in 2010–11 to 90% metering by 2029–30 (Ofwat, 2011). The review noted the potential for smart metering but did not specifically consider it. In 2012, Ofwat maintained that it is not concerned about what technology is used by each company (Ofwat, 2012). It is the output that matters. DEFRA believes that there is a 'perception' that water utilities are not innovative because of their regulatory framework. 'Not getting in the way' ought to be the government's main objective (Phippard, 2012). Smart water appears to remain a low priority, with Ofwat only referring to smart metering for commercial customers in its March 2012 position paper on innovation in the sector (Ofwat, 2012).

The smart meter deployment plans in AMPs 5, 6 and 7 by Southern Water and Thames Water (Chapter 4) were decisions made by the companies alone and Ofwat has remained neutral on the subject of smart metering. There are no policies that in fact impede plans to develop smart water network at the national level.

Some progress has since been seen. Ministerial support in 2013 for combined sewer outfall monitoring by 2020 (Chapter 5.6.5) is being used to develop smart sewerage monitoring (Hulme, 2015). In 2017, Ofwat unveiled a strategy for companies to use customer data from 2020 to improve customer service and communications. While not alluding to smart data management, the report highlights the potential for an appreciably greater depth of data gathering and integration. Areas highlighted include using data from credit rating agencies to identify financially vulnerable customers improve its debt management and customer support (developed by Yorkshire Water), using postcode hotspots to target customers who may be flushing excess fat and grease into the sewerage networks (Southern Water), data analytics for understanding customer behaviour (Dŵr Cymru Welsh Water and Affinity Water) and peer-comparison for customer water consumption (South East Water with Advizzo). This is the first position paper to state the advantages of customer data analytics (Ofwat, 2017).

Scottish Water – competition as a driver: Scottish Water is a state-owned and operated company that operates within a similar regulatory framework to the privatised companies in England and Wales being regulated by the Water Industry Commission for Scotland (WICS). Competition for all non-domestic customers was introduced on 1 April 2008, under the Water Services etc. (Scotland) Act 2005. This affects all of the utility's 130,000 non-domestic customers. 45,000 contracts have been renegotiated in the first two years since the act was passed. Scottish Water has announced that it will install up to 3,000 smart meters in public buildings throughout Scotland, largely based on a business case argument that water savings through better leakage control and reduced consumption will benefit both the company and its customers (Staddon, 2012; WICS, 2010).

Scottish Water has used its Business Stream customer relationships to deliver a dedicated customer service when the large user market was opened up. By pre-empting customer concerns and delivering customer-centred service prior to competition, switching has been quite modest. Between 2010–11 and 2020–21, the act is forecast to generate savings of £60–70 million from lower unit prices and £50–55 million from lower water use. Actual savings have been greater; in the eight years since its launch, initiatives by Business Stream have resulted in customer bills being lowered by £99 million through lower charges, along with £53 million saved through reduced water consumption (24 million m^3 of water) and £7 million in lower electricity use. Total savings were £37 million in 2015–16 alone (Gaines, 2016).

Examples of savings through smart metering include an 8% reduction of water use by Tesco stores in Scotland (Ofwat, 2012). Smart meters installed at NHS Lothian's Royal Victoria Hospital identified anomalous water consumption patterns which resulted in a hitherto undetected leak being repaired, lowering water consumption from 12 m^3 per hour to 4 m^3 per hour, saving £130,000 per annum (Business Stream, 2016a). Smart water meters were used at News UK's printing press at Eurocentral, Lothian to identify areas to lower consumption, including installing water-limiting taps and efficient shower-heads and replacing faulty lavatory cisterns (Business

Stream, 2016b). With water consumption decreasing by 50%, the water bill was lowered by £5,000 per annum and fixed costs reduced by £3,000 per annum through a smaller meter size subsequently being needed.

In its 2010 review of the Act's progress, WICS does not mention smart water. The deployment of smart metering has been a response to policy, not directly seeking to encourage metering usage (WICS, 2010), let alone smart metering.

8.6.7   Case Study 8.7: The USA – State Level Mandates

In the USA policy is developed and enacted at the state and local level. The most significant policy objectives have been those associated with water resources management and demand reduction at the state level.

Supporting initiatives in California and Arizona: In California, Senate Bill SB ×7-7 Delta & Water Reform Legislation of 2009 mandates the state to reduce urban per capita water use by 20% by the end of 2020, along with end of 2015 interim use targets, as defined by each municipality. In April 2015, a new 25% reduction target was set, including mandatory drip landscape irrigation for all new homes (Gambino, 2015). For the utility, it is easier to drive down consumption by eliminating data errors (missing or out-of-calibration meters, incorrect meter data, etc.), system losses and water loss through theft in order to get a true understanding of the their non-revenue water, rather than undertaking wholesale pipe replacement. Under California's Proposition 218 (1996) a water rate increase must be based on the cost of providing the service. This effectively prohibits local governments from using cross subsidies between for example irrigation and residential customers. In order to demonstrate that this 'proportionality' is consistent, more precise data is required, which has been seen to support the installations of smart grid technologies. Mandatory water restrictions were lifted in May 2016, but the long-term water use reduction targets remain in place. As a result, a number of towns and cities are rolling out smart water meter programmes (M&SEI, 2016). Also in May 2016, the state allowed utilities to develop standards that met their particular needs, also reflecting that changing water needs and responses to these can outpace formal policy development (Freedman and Dietz, 2017).

Arizona's Department of Water Resources (ADWR) Modified Non-per Capita Conservation Program was developed between 2006 and 2008 and implemented in 2010. The programme requires water providers to adopt best management practices to achieve water conservation. While this does not specifically require advanced metering, it lays down a framework for future smart water implementation. In Arizona a survey of water conservation measures in 2010 found that five of 15 communities offered rebates relating to smart irrigation. These included rebates of $22–100 for irrigation audits and rebates of $30–250 for domestic lawn smart irrigation systems and in one case $5,000 (or one third of total cost) for commercial smart irrigation upgrades. Smart metering was not covered by the survey (Western Resources Advocates, 2010). Since 2013, each July the Arizona Municipal Water Users Association has been the state's 'smart irrigation month' when domestic users are encouraged to audit their garden irrigation and to install or upgrade their smart irrigation systems.

Subsidies for smart irrigation: AquaSpy have noted that some local government entities such as the NRCS in Texas and the USDA in Georgia have subsidised the purchase of the company's data subscription services by 50% and therefore are making it more attractive for growers to consider (Moeller, 2012, personal communication). However, they believe that the general picture is of an absence of coordinated and significant programmes helping drive demand nor standards, nor even awareness about these.

Mandating smart municipal irrigation: This also applies for amenity irrigation and in California and Texas, where parks, gardens and sports fields are mandated to adopt smart irrigation (OECD, 2012). In Texas, the HB 2299 bill relating to equipment used for irrigation systems came into effect in 2008. The bill calls for irrigation devices to be fitted with smart controllers and water and was modified to call for automatic shut-off systems for periods of rain and frost from 2011. In 2007, five types of smart controllers were being sold in the state which rose to 11 by 2011 (Lee, 2011). California's Updated Model Water Efficient Landscape Ordinance AB1881 states that from 2012, all new irrigation devices in the state must have smart controllers. In 2015, the Texas Irrigation Association followed Arizona's lead and declared July as the state's 'smart irrigation month.'

Federal Government initiatives are emerging: The 2015 Smart Energy and Water Efficiency Act (HR 3413) includes supporting smart water grid pilot projects in three to five cities with an emphasis on leak detection and remediation. That year, The US National Institute of Standards and Technology supported a project where AT&T and IBM have been working with the cities of Los Angeles, Las Vegas and Atlanta to develop a smart leak management network under its NIST Global Team Challenge and the US Bureau of Reclamation also provided a $1 million grant to support smart water metering at the Gateway Water management Authority in California (Freedman and Dietz, 2017).

Conclusions

This chapter is in part drawn from the author's examination of smart water and policy (OECD, 2012) in 2011–12. What has changed since? Progress in policy development appears to have been limited in recent years. This reflects a continuing tendency to attach a low priority to water management at government level.

Even so, some developments have been seen. In Singapore, there has been a shift towards appreciating the value derived from household level smart metering. Until 2017, there has been piecemeal progress in the same direction in parts of England. Ofwat's 2017 position paper on the use of analytics for customer-related data from 2020 indicates a new willingness to consider innovative approaches. In Korea, the national smart grid of smart grids by 2020 remains a national priority and the essential infrastructure is being trailed. Singapore is focusing its various smart elements towards a more unified outcome.

One particular concern is that the pace of technological development for smart water applications is outpacing the capacity of governmental institutions to respond to these (Freedman and Dietz, 2017). Policy initiatives may need to be devolved to where such innovations are emerging so that they can be accepted and incorporated where and when they prove to be of a public benefit.

Despite droughts abating in Australia and the USA, the regulatory desire to improve water efficiency has been maintained and with this, the adoption of smart responses continues to develop.

It would seem that smart water development and deployment will in effect continue to be driven by indirect policy influences. This may change as the size and profile of smart water markets grow. It cannot be assumed that a higher profile will inevitably be beneficial. As will be discussed in Chapter 9, there are a number of challenges to be faced before smart water management becomes more widely adopted.

References

ACT (2007) Water Resources, (Amounts of water reasonable for uses guidelines) Determination 2007 (No 1). Australian Capital Territory, Canberra, Australia.

Business Stream (2016a) A leak sees money trickle away. Business Stream, Scottish Water.

Business Stream (2016b) When small changes can make big savings. Business Stream, Scottish Water.

Byeon S, Choi G, Maeng S and Gourbesville P (2015) Sustainable Water Distribution Strategy with Smart Water Grid. Sustainability (7): 4240–4259.

Brzozowski C (2011) The 'Smart' Water Grid: A new way to describe the relationship between technology, resource management, and sustainable water infrastructures. Water Efficiency 6 (5): 10–23.

Choi H and Kim J A (2011) Alternative Water Resources and Future Perspectives of Korea. 2011 IWA-ASPIRE Smart water Workshop, 4th October 2011, Tokyo, Japan.

Choi G W, Chong K Y, Kim S J, and Ryu T S (2016) SWMI: new paradigm of water resources management for SDGs. Smart Water 1: 3.

Cleantech Group (2011) Ontario Global Water Leadership Summit. Cleantech Group, San Francisco, USA.

Fisher S (2016) Addressing the water leak challenge in Copenhagen. WWi June–July 2016, 32–33.

Freedman J and Dietz G (2017) The Future of Water Management: A Menu for Policymakers in the Digital Industrial Era. The General Electric Corporation, Boston, MA, USA.

Gaines M (2016) Business Stream saves customers £37m. Utility Week, 5th December 2016.

Gambino L (2015) California restricts water as snowpack survey finds 'no show whatsoever'. The Guardian, 1st April 2015.

Hulme P (2015) The Need for EDM. Presentation to 'The Value of Intelligence in the Wastewater Network', CIWEM, London, 18th February 2015.

ITU (2010) ICT as an Enabler for Smart Water Management. ITU-T Technology Watch Report, ITU Geneva, Switzerland.

Lee L (2011) The art of smart irrigation. Tx:H2O 6 (3): 10–12.

Lloyd Owen D A (2012) Singapore – a holistic approach to sustainability. Paper presented to the Singapore International water Week, July 2011.

Luger J (2011) Key developments in Smart Metering. Ofwat, Birmingham, UK.

MacDonald K (2010) Recent leakage performance and future challenges. Presentation at the SBWWI Metering and Leakage Seminar – New Challenges: New Solutions, 24th November 2010, Leamington Spa, United Kingdom.

M&SEI (2016) Analysis: California smart water meter landscape. Meter & Smart Energy International, 9th August 2016.

Merks C, Shepherd M, Fantozi M and Lambert A (2017) NRW as a percentage of System Input Volume just doesn't work! Presentation to the IWA Efficient Urban Water Management Specialist Group, Bath, UK, 18–20th June 2017.

OECD (2012) Policies to support smart water systems: Lessons from countries experience. ENV/EPOC/WPBWE(2016)6, OECD, Paris, France.

Ofwat (2010) Playing our part – reducing greenhouse gas emissions in the water and sewerage sectors. Ofwat, Birmingham, UK.

Ofwat (2011) Exploring the costs and benefits of faster, more systematic water metering in England and Wales.

Ofwat (2012) Valuing every drop – how can we encourage efficiency and innovation in water supply? Ofwat, Birmingham, UK.

Ofwat (2017) Unlocking the value in customer data: a report for water companies in England and Wales. Ofwat, Birmingham, UK.

Phippard S (2012) Comments at the World Water-Tech Investment Summit, London, February 2012. Director, Water, Floods, Environmental Risk & Regulation, DEFRA.

Public Utilities Board Singapore (2016) Managing the water distribution network with a Smart Water Grid. Smart Water 1: 4.

Renner S, et al. (2011) European Smart Metering Landscape report. Intelligent Energy Europe, Vienna, Austria.

Smart Approved WaterMark (2011) Final report on the delivery of the Smart Approved WaterMark Water Smart Australia, Project Report to the Department of Sustainability, Environment, Water, Population and Communities. SAWM, Sydney, 2011.

Staddon C (2012) Department of Geography and Environmental Management, University of the West of England, Bristol, Personal communication, 2012.

Turner A, et al. (2010) Third Party Evaluation of Wide Bay Water Smart Metering and Sustainable Water Pricing Initiative Project. Report prepared by the Snowy Mountains Engineering Corporation in association with the Institute for Sustainable Futures, UTS, for the Department of the Environment, Water, Heritage and the Arts, Canberra.

Western Resources Advocates (2010) Arizona Water Meter: A Comparison of Water Conservation Programs in 15 Arizona Communities. WRA, Boulder, Co, USA.

WICS (2010) Competition in the Scottish water industry 2009–10. Water Industry Commission for Scotland.

Worsfold M (2012) Head of Asset Strategy, Ofwat. Comments at the World Water-Tech Investment Summit, London, February 2012.

9

Obstacles to Adoption

Introduction

Using disruptive technologies, as implied, involves at least a degree of disruption. Change in an essentially conservative service such as water and sewerage will not necessarily enjoy swift or universal acceptance.

Data privacy laws in the UK, Netherlands and California USA, and fears about possible health implications of data transmission, have been used to postpone or even prevent the installation of smart water meters. Concerns about stranded assets can also hinder the deployment of smart metering, where recently installed assets would have to be replaced and likewise, innovations have employment implications. Other challenges include the need to ensure the effective (and therefore beneficial) adoption of new technologies, maintaining the integrity of systems, questions about operating and transmission standards and interoperability along with the ownership of the data being generated.

9.1 Public Concerns about Health and Privacy

In the Netherlands smart metering systems for power grids and natural gas will be implemented, according to European law (2012/148/EU) by 2020, and it was originally assumed that this would be used as a platform to install smart water meters over the same time. The Netherlands traditionally assumed that long-term water policy objectives overrode single-issue concerns. The 'greater good' does not always apply; in 2010, the Netherlands had to postpone its universal smart water metering policy because of media and 'vocal minority' opposition, relating to concerns about health and privacy. Due to their tradition of not taking these concerns into account, many utilities found that they were unable to respond constructively to such unexpected opposition.

Smart water deployment is hindered when policies limit the collection of data from consumers, for example regarding privacy laws and data security requirements. The California Public Utilities Commission has allowed customers to opt-out from using smart meters over concerns about electromagnetic fields (EMFs) generated by the meter radio transmitter. This reduces the utility of the 'smart' meters and drives up costs as customers have to have their meters manually read.

In the UK, similar concerns have been expressed regarding the Data Protection Act and the amount of personal information smart meters can obtain (Harper, 2011). This

Smart Water Technologies and Techniques: Data Capture and Analysis for Sustainable Water Management,
First Edition. David A. Lloyd Owen.
© 2018 John Wiley & Sons Ltd. Published 2018 by John Wiley & Sons Ltd.

places an emphasis on data security (what if a malevolent entity gains control over water devices, in networks or in homes?) and the way the data is processed. With legitimate reasons for processing data, at an appropriate collection frequency and customer access to this data, it is legal (Murray, 2011). In the UK, newspaper headlines such as, 'My water meter can be a killer' (Williams, 2010) and 'Not so smart meters will "enable snooping and pose a health risk"' (Casey, 2016) demonstrate the sensitivity of policy to objections that are unlikely to have been rationally anticipated.

Websites such as 'smartmetermurder.com' and 'stopsmartmeters.org' in the USA and UK allege that smart meters damage the users' DNA, cause cancers, kill local wildlife and so on. In the current populist climate, such points of view can acquire more traction than had been previously appreciated.

These concerns are to some degree common to smart networks in general. Such concerns (OECD, 2012) are not easily alleviated without consultation and appropriate safeguards that empower consumers. Smart applications can potentially give insight into many details of people's lives and security breaches are an ever-present risk. They all potentially provide personal information to third parties as well as offering better services for consumers (Murray, 2011). In the UK (Hall, 2014), this concerns legislation including the Data Protection Act (1998); Human Rights Act (1998); Privacy and Electronic Communication Regulations (PECR).

Another potential privacy concern may arise from the ability of smart sewer networks to detect the consumption of drugs, both legal and illegal. The smart lavatories noted in Chapter 4 have the capability of collecting drug related data at the user level (Ratti et al., 2014) and transmitting this to third parties as do the smart sewers seen in Chapter 5. In theory at least, sewerage data could be used to track people worldwide, wherever the necessary probes are in use and are connected to the data needed, assuming the personal DNA data is also available, which could mean that for the right or the wrong reasons a new way of monitoring people's whereabouts could be unveiled.

This goes back to some of the libertarian arguments against smart water meters. For example, an employer could soon have effective data on its staff's personal habits. It could have this without the staff knowing, let alone their consent. Such applications are limited only by what is contained within water and wastewater flows and what can potentially be deduced from them.

9.2 Trust, Technology and Politics

A lack of confidence in the potability of water supplies, even where they are of a high quality has resulted in the increasing adoption of non-utility water sources, including point of use treatment (PoU) systems (Gasson, 2017). One of the drivers behind this is the fact that improved detection and monitoring means that contamination is being found where it could not previously be detected, as is the case where smart water monitoring systems are being deployed. This is the paradox of cleaner water being regarded as getting dirtier as detection limits move from parts per million to parts per billion.

Responding to concerns about trust can either be approached through improved real-time monitoring and making this information more freely available or through localised monitoring. The technology needed for PoU monitoring is currently too expensive for wide adoption, but as probes become cheaper, this is likely to emerge as will be discussed in Chapter 10. Trust and confidence also depends on the country. In England and Wales, the Drinking Water Inspectorate (dwi.gov.uk) is independent but operates on a statutory basis. It is held in high regard by utilities and consumers, with 99.96% of drinking water samples meting all health and aesthetic criteria in 2015 (DWI, 2016) against 98.65% in 1992 (DWI, 2002).

Innovation brings its own challenges. Without adequate training and preparation, installing smart meters can result in false leak alerts and unnecessary network repairs. This undermines confidence in a new technology, which ought to have been addressed at the outset (Cespedes and Peleg, 2017). Utilities need to be able to effectively communicate the benefits of smart networks. This means considering what customers actually need to know and getting the relevant messages across, for example, informing customers about how rapidly leaks are repaired, dividing data into utility sub-districts, and having weekly and monthly performance goals to motivate staff to operate more effectively (Cespedes and Peleg, 2017). In Chapter 4, approaches towards building customer relationships by Southern Water and Thames Water were reviewed.

9.3 Ownership of Data

Ownership of data is an emerging area of concern. Companies developing smart hardware and software are able to gather data from companies using their offerings and monetise these. For some water metering services, this is used as a tool for helping to influence consumer behaviour through peer-group benchmarking. Here, the objection may be one of perspective, when people provide data freely which is then used as a commercial tool.

In other cases, for example Fathom (domestic smart metering), WeatherTrak (weather data) and AquaSpy (soil moisture), the gathering of data for benchmarking is what subscribers are paying for. Data may also be used by a company as a marketing tool. For example, if one company has a range of smart water tools all with common interfaces and operating systems, it can become a way of building customer loyalty through these common platforms.

9.4 Stranded Assets

Municipally owned water utilities in the Netherlands appear to be divided about smart water meters. Some companies like Vitens, WML and PWN see opportunities in the drinking water sector and will install smart water meters in the coming years. They install a smart water meter free of charge, and see this as an opportunity to serve customers better. Amsterdam's Waternet will not install these meters since the company only recently started installing new water meters, and replacing these meters with smart meters is seen as a waste of money (Struker and Havekes, 2012). This is an example of the challenge of stranded assets, where installing a new technology involves replacing hardware that is fully operational, albeit having been left behind by subsequent developments.

9.5   The Role of Utilities

Utilities need regulatory and operational incentives to adopt smart water approaches. There is a substantial lobbying force that maintains that municipal water should be 'free', either paid indirectly through taxation as in Ireland or somehow cross-subsidised by industrial customers (Barlow and Clarke, 2002; Barlow, 2007). If utilities are unable to charge for water consumption, no incentive for demand management exists. Where utilities operate on a cost pass-on principle as in much of the USA, proactive leak management does not help the utility, as they are incentivised by regulators and the media to repair leaks rather than to prevent them or to repair before they are noticed (Cespedes and Peleg, 2017). Prevention neither creates work nor does it show a job being done in public.

Another concern is that utilities should not have any form of involvement with the private sector (Hall and Lobina, 2007) irrespective of any gains such partnerships may deliver. One of the reasons for this is in order to protect jobs. This is relevant to smart water, as one of its intended outcomes is to minimise operating and capital spending through optimising efficiency. One particular area here is meter reading. Automating meter reading will result in job losses. In developing economies, service extension to date has seen new jobs replace old ones; a water vendor (original job, non-utility) can become a bill collector (original job, now with a utility) and in turn a meter reader (new job, same utility). Utilities need to consider the social implications of AMR and AMI for their staff.

9.6   Integrity and the Internet

There are two aspects to smart water data and confidentiality. Gaining access to remote water meter transmission data could become a way of finding out if a building is occupied or empty by observing water usage patterns. This discounts the fact that electricity meters operate in the same manner, as do telephone lines and indeed do all Internet-based devices. In addition, as with other timed devices, water consumption is not necessarily a sign that the occupier is in or out at the moment in question.

In Chapter 10, the potential for smart water as a sub-set of the Internet of Things (IoT) will be considered. One of the underlying principles of the IoT is that it makes a property 'smart' through interconnecting all applicable devices, including those that are involved with water. As this involves a large number of interconnected devices, the most vulnerable smart device may allow access to other devices and customer account data and settings. Cyber security is a potential concern when a large number of individual devices need suitably robust passwords. A variety of low-cost devices may be installed in a household without consideration (or indeed realisation) about the need for password protection. Any single device within such a network could be a point of entry to the rest of the network.

9.7   A Question of Standards Revisited

The International Telecommunications Union (ITU, 2010) notes that it has developed common standards for the Internet of Things ubiquitous sensor networks; such standards have yet to be drawn up and applied in smart water management (UNESCO & ITU, 2014). The principal concern here lies in the interoperability of the various sub-units. As a typical smart water system relies upon a significant number of data sources and processes any erosion in efficiency caused by components that are not fully inter-operable may have a cumulative and deleterious effect. This is the counter-side to the potential for smart systems to bring together incremental improvements into a disruptive whole.

A lack of integration between smart water systems may also hold back further gains through the sharing of common data, such as weather-related flooding in different river basins or aspects of consumer behaviour across a number of water utilities, or even within a utility.

This also applies with the data collection, data processing and presentation aspects of smart water management. Generally speaking, data communications and transmission work within the extensive sets of standards and protocols that have already been implemented by the ITU for telecommunication and electricity, but not nearly to the same extent for water. In Chapter 8, ISO standards that will form the basis for smart water systems were outlined.

Parallel to this is the need for common data standards. For example, there is the preference in the USA for water utilities and regulators to use imperial measures, while the metric system is the norm in most of the rest of the world. As each catchment area has a unique combination of physical and chemical characteristics utilities need to reconcile these when adopting a smart water system in another catchment area. This has been a long-standing concern in, for example, the United Kingdom, when innovators seek to have pilot projects carried out in one utility accepted by another utility.

9.8 Demand Management and Flushing Sewage Through the Network

In Germany, efforts to reduce domestic water consumption have caused some problems with the effective flushing of sewage through the foul sewer network, including increased residence times of solids before they reach the sewage works. Average domestic consumption fell from 144 l/cap/day in 1991 to 124 in 2011 (Gersmann, 2012), and in order to deal with fecal contamination build-ups in the sewer networks, increased chemical treatment is being used in cities such as Cologne which cannot now use gravity-based sewer flushing (Chandavarkar, 2009). This needs to be put into the context of Copenhagen (Chapter 5.8.1) where no sewerage issues emerged from the reduction of per capita water consumption from 170 l/cap/day to 100 l/cap/day.

9.9 Data Handling Capacity for the Internet of Things

The Internet of Things will require a communications infrastructure capable of carrying the increased volume of data involved. The 5G (fifth generation) mobile communications standard is been developed for data intensive applications (5G PPP, 2016). Local trials of 5G systems have been taking place since 2016, with the aim of a formal global standard being adopted in 2018 (Woods, 2017) with launches in China and Korea anticipated in 2019 (Fildes, 2017). Its effective roll-out in the UK and other countries is not anticipated before 2020 and broader adoption is unlikely in most countries before 2025 at the earliest due to the limited capacities of current networks.

One particular challenge will be developing a suitable communications infrastructure. In urban areas, a dense network of communications cells will serve relatively small areas. A survey of the literature associated with 5G and utilities by the author indicates that water management has to date been one of the more peripheral areas of concern.

9.10   Leakage Management is Hampered by its Measurement

No single leakage indicator exists for all purposes, and so measures need to be applicable for particular issues (European Commission, 2015). The Chartered Institute of Water and Environmental Management notes that leakage 'should definitely not be quoted in terms of percentages of system input volume; it is misleading for comparisons because of differences and changes in consumption, and it is a zero-sum calculation which cannot identify true reductions in leakage and consumption in the same time period' (CIWEM, 2016).

Unfortunately, as noted by Aguas de Cascais (Portugal) 'we have to put up with' percentage NRW and leakage data as this is what the customer, along with politicians and the media understand. While ILI, litres per capita and litres per meter data is more objective, it is not so easily comprehensible to the non-specialist user, whatever its failings (Perdiago, 2015).

Measures such as Ml/day and litres/property are traditionally have been used effectively in the United Kingdom, along with m^3/km of mains for more rural areas. Infrastructure Leakage Index (ILI) used in conjunction with some measure of pressure is more reliable for international comparisons of technical performance. As discussed in Chapter 8.3, some headway has been made here, but the compelling simplicity of a percentage measure, no matter how misplaced, will take some shifting.

9.11   Smart Water has its Logical Limits

Is there a point where the application of 'smart' ideas to water goes too far? This is perhaps summed up by Bradshaw (2015) when considering the point of 'smart' drinking vessels, which measure how much liquid you drink and relate this to your recommended daily intake. These require the user to decant everything they drink into a 'smart' plastic cup that sends water consumption data to your smartphone. Smart water ought to be about efficiency and minimising the amount of equipment we need to manage our water use.

Conclusions

Many of the obstacles that face the deployment of smart water approaches reflect the challenges faces by those seeking to offer innovations in the water sector in general.

While industry and agriculture (irrigation) are broadly pragmatic, it is fair to say that much of the utility side is to some degree more dogmatic. It is also necessary to appreciate that no two water utilities are the same in terms of their water resources and requirements, and due to their fragmented nature, utilities tend to look for different responses to the same circumstances.

The potential pitfalls outlined above are those that the author is aware of. There may well be many others. Some may be trivial, but others carry the risk of delaying the deployment of smart approaches for a decade, if not more. Obstacles to innovation can be characterised by their uncertain or irrational nature.

Ambitions tend to encounter realities at some point. Such encounters do not have to result in setbacks. Innovations developed in isolation may have a limited use compared with those which have enjoyed a degree of interaction with potential adopters.

Do innovators actually know what utilities and other customers want? This can be a problem especially where smart water applications have emerged from offerings developed for other sectors. There is a need for effective communications between developers and utilities and between utilities and their customers. In essence, there are two great challenges: to ensure that systems and components are interoperable; and to ensure that people are fully engaged with the process of innovation so that they are best placed to benefit from it.

References

5G PPP (2016) 5G empowering vertical industries. 5G PPP, Ghent, Belgium.

Barlow M and Clarke T (2002) Blue Gold: The Fight to Stop the Corporate Theft of the World's Water. The New Press, NY, USA.

Barlow M (2007) Blue Covenant: The Fight for Water as a Human Right. McClelland & Stewart, Toronto, Canada.

Bradshaw T (2015) 'Smart vessels' quench a non-existent thirst. Financial Times, 19th November 2015.

Casey C (2016) 'Not so smart meters will "enable snooping and pose a health risk"'. Enfield Gazette & Advertiser, 23rd November 2016.

Cespedes F V and Peleg A (2017) How the Water Industry Learned to Embrace Data. Harvard Business Review, Digital Article, 27th March 2017.

Chandavarkar P (2009) German water conservation impairs sewage treatment. Deutsche Welle, 20th August 2009.

CIWEM (2016) Water distribution system leakage in the UK. Policy position statement, June 2016, CIWEM, London, UK.

DWI (2016) Drinking Water 2015. Drinking Water Inspectorate, London, UK.

DWI (2002) Drinking Water 2001. Drinking Water Inspectorate, London, UK.

European Commission (2015) EU Reference document Good Practices on Leakage Management WFD CIS WG PoM. European Commission, Brussels, Belgium.

Fildes N (2017) UK sets timetable for launch of 5G networks. Financial Times, 8th February 2017.

Gasson C (2017) A new model for water access. Global Water Leaders Group, Oxford, UK.

Gersmann H (2012) Germany's careful toilet-flushing is a drop in the water-conservation ocean. The Guardian, 18th April 2012.

GWI (2016) Chart of the month: Utility vs discretionary spending. Global Water Intelligence, 17 (11): 5.

Hall D and Lobina E (2007) Water as a public service. PSIRU, Business School, University of Greenwich.

Hall M (2014) Pioneering Smart Water in the UK. SMI Smart Water Systems Conference, London, April 28–29th 2014.

Harper N (2011) AMR – Does it do what it says on the tin? SBWWI 23rd November 2010.

ITU (2010) ICT as an Enabler for Smart Water Management. ITU-T Technology Watch Report, ITU Geneva, Switzerland.

Murray S (2011) Dealing with data. SBWWI, 6th December 2011.

Perdiago P (2015) A smart NRW reduction strategy. Presentation to the SMi Smart Water Systems Conference, London, April 29–30th 2015.

Ratti C, Turgeman T and Alm E (2014) Smart toilets and sewer sensors are coming. Wired, March 2014.

Struker A and Havekes M (2012) Waternet, Personal communication.

UNESCO & ITU (2014) Partnering for solutions: ICTs in Smart Water Management. ITU-T UNESCO/ITU, Geneva, Switzerland.

Williams L (2010) 'My water meter can be a killer.' Liverpool Echo, 9th June 2010; The Daily Mail, 9th June 2010.

Woods B (2017) What is 5G and when will it launch in the UK? Wired, 7th June 2017.

10

Towards Smart Water Management

Introduction

This chapter draws together some of the themes explored in this survey and considers how they may be able to deliver further improvements through their effective integration. As early stage innovations have been examined through this book, it concentrates on products and services that have already been developed, rather than ones that may emerge at some point in the future.

10.1 Conservatism and Innovation

The conservative approach towards accepting and adopting innovative technology by water utilities means that unless a system is currently being developed and trialled, it may be assumed that they will not be broadly used before the 2030s (Sedlack, 2016).

This is in contrast to the development to deployment trajectory seen in 'smart' technologies such as mobile communications and computing, as highlighted by the smartphone. The Nokia 9000 Communicator (a mobile phone with an integrated personal digital assistant or PDA) was launched in 1996 (Nokia, 1996). Two years after its introduction in Japan in 1999, the NTT DoCoMo achieved mass adoption (Anwar, 2002) and between 2003 and 2006, fully integrated smartphones such as the Blackberry series were adopted internationally and in 2007 Apple launched the iPhone. By 2013, smartphone sales outstripped those of conventional mobile handsets (Gartner, 2014). Here what is noticeable is not just the speed of their adoption, but the evolution of the devices themselves within this time.

The smartphone analogy is a useful one; Apple's business model demonstrates how consumers may find a new offering indispensable despite being quite unaware of its existence a relatively brief time before. That is the ideal in developing and marketing new products and finding new markets for them. Meeting previously unrealised needs is beneficial when it offers genuine utility. This approach lies at the heart of many smart enabled devices and applications. Innovation can be driven by developers devising applications (and apps) through their personal experience to address needs that have hitherto been overlooked.

Smart Water Technologies and Techniques: Data Capture and Analysis for Sustainable Water Management, First Edition. David A. Lloyd Owen.
© 2018 John Wiley & Sons Ltd. Published 2018 by John Wiley & Sons Ltd.

Water, in contrast, has been, and to a large extent remains, a business characterised by its risk-averse outlook. That will remain the case unless consumer and other stakeholder expectations can be transformed by innovations that may change their relationships and expectations. For the consumer, the readiness to accept and adopt changes to some extent depends upon raising their expectations from that of passive acceptors of a standard service, into engaged participants able both to modify their consumption patterns and to influence broader water management policy.

Other influences include the increased need to meet stakeholder expectations regarding environmental compliance, service delivery and public health obligations. Such scrutiny may not be welcome. In the first half of 2017, Thames Water accrued a net £35.5 million in fines and penalties, including a £20.3 million fine for the illegal discharge of 4.2 million m^3 of sewage at six outflows (Prescott, 2017) and regulatory penalties including £8.6 million for failing to reach its leakage management target (Thames Water, 2017). The Chairman of Ofwat (Cox, 2017) subsequently called for extensive changes to the company's governance and reporting systems, and has prioritised a more aggressive approach to leakage reduction across the sector for 2020–25 in consequence (Ofwat, 2017). The need to meet these challenges may make utilities more open to innovation.

What cannot be transformed is the fact that this involves dealing with low-value, high-volume substances which are not usually be regarded as a commodity. The delivery of clean water and the safe removal of sewage do not fit in easily with the public perception of smart futures based upon deriving the greatest value from the smallest entities. For this reason, industrial customers have an important role in assisting early stage innovations to reach commercial maturity. The industrial customer is characterised by its need to use water as efficiently as is possible in order to lower its operating costs (less water abstracted and consumed, fewer effluents generated and lower power and chemical consumption), while complying with applicable environmental and public health standards. An industrial customer will adopt innovations where it makes business sense to do so. Companies seeking to develop and commercialise their smart water offerings need to consider the potential for industrial customers as a bridge towards utility markets.

Water utilities can respond to innovations when it is in their interest to do so. 'Wastewater is an oxymoron' (Parker, personal communication, 2011) and over the past decade there has been a concerted move towards water reuse and energy and nutrient recovery, creating commodities from what was previously seen as a waste. For example, since the end of 2016, energy recovered from sewage in Aarhus (energy generated is 150% of the energy used during treatment) is used to power water distribution and treatment, making the utility a net generator of energy (Karath, 2016).

10.2 A Set of Desired Outcomes

Long-term forecasts or idealised visions as to what smart water systems may appear as tend towards two outcomes: what might be desired and what may realistically be achieved. The following suggestions are the author's subjective set of desired outcomes for smart water by 2030. This is also aims to synthesise the various products and services that have been described in the previous chapters and to outline the potential for their broader integration.

Water abstraction and treatment: Integrated realtime monitoring of water flow and quality throughout the catchment area to the point of abstraction and into the bulk water network to the water treatment works. The treatment works is optimised to anticipate any changes to the raw water before it enters the facility. The treatment works is also linked to current and forecast demand for treated water to ensure the minimum amount necessary is treated and resides within the distribution networks. Catchment level and abstraction point data is further integrated into a national water grid.

Water distribution: Pressure, flow and quality monitoring through the distribution network to ensure minimal pumping is used, thereby minimising unavoidable leakage and detecting actual or incipient leakage at or near to the point of occurrence. Monitoring of water quality indicators highlights potential pipeline deterioration or biofilm accumulation.

Leakage management: Integrating distribution network flow and loss data with external monitoring such as satellite or drones for the presence of potable (treated) water outside the pipes through its spectral 'signature' in the ambient environment.

Demand management: Realtime advice about domestic water usage and its water and energy and cost implications available by a personal device, along with realtime alerts about anomalous water consumption and the ability to remotely shut water off in the event of a leak. Integrated smart metering would in turn provide more sophisticated customers with data on their water use, recycled water use and wastewater and drainage water discharge along with water-related electricity use.

Customer management: Address affordability concerns through the integration of customer water use and their personal circumstances, and to align these with tariff packages (including social tariffs) and demand management to avoid bad debts accruing.

Effluent collection: Remote monitoring of soil temperature for sewer leak alerts. Effluent flow monitoring and pollution load detection related to network capacity and any potential storm water. Changes in effluent temperature within the pipe network monitored to measure extent of interactions between foul and storm sewerage.

Effluent treatment and resource recovery: Monitoring of current and potential flows of foul water and storm water to ensure adequate treatment capacity is available. Using effluent and stormwater holding areas to even out effluent flow and match it with treatment capacity. Realtime or near realtime monitoring of the treatment process to ensure optimum process efficiency is maintained along with post-treatment discharges being in the best state for water, energy and nutrient recovery. Discharge data from the works and all flow systems related to it are provided in realtime to all relevant third parties.

Public health: Presence of pathogens or genetic tracers associated with them in urban sewerage systems used to pinpoint outbreaks and spread of infectious diseases and to alert about any mutations in these organisms. Abnormal levels of recreational or medicinal drugs or their metabolites have the potential to warn about other emerging concerns. By detecting at the sewer rather than the lavatory level, personal privacy concerns can be addressed. Gene sequencing can be used for rapid disease outbreak diagnostics and a mobile alert to the presence of infection and record its spread. Minimising the time for disease notification means there is more time to respond and contain its spread.

Flood resilience: A realtime river basin level understanding of current and predicted river and surface water flows to provide the greatest amount of warning of future flood events. Current and historic data is continually fed into the system to improve the accuracy and adaptability of the system and to take into account any physical changes in the area, along with testing for hitherto unforeseen rainfall levels. This is continually interlinked with measures being taken to improve flood resilience and storm water storage capacity within the river basin.

Inland water quality: Sources of diffuse pollution can be detected through gene sequence mapping and identification for the rapid detection of bacterial contamination and its location. This starts with identifying the species (human or cattle feces, for example) involved before working backwards to its source using marker pellets with unique DNA sequences.

River basin water resources management: This includes the integrated monitoring of water use against resources in order to pinpoint current and potential use conflicts. This can also identify where water pricing elements are out of line with its availability and use.

Integrated decision-making: Managing a sustainable balance between various user stakeholders (municipal, industrial and agricultural users) and maintaining the environmental integrity of the inland water systems.

Irrigation: Integration of yields from irrigated areas to the amount of water used to benchmark their efficiency. At the river basin level, the ability to predict water demand for agriculture, industry and municipalities and to take timely action to obviate potential conflicts between them.

Industrial customers: Providing data about the efficiency of water use and recovery and to benchmark it to best practice within the sector and the river basin. This would in turn allow for the generation of usage and compliance data for third parties to provide reassurance about the impact of the company's activities on local water resources and inland water quality. Where feasible, smart monitoring of water usage and recovery to minimise or indeed eliminate net abstraction from surrounding water sources.

Developing economies: Create a realisable and affordable pathway to developed economy standards of water and sanitation access through the effective deployment of intermediate approaches. These include networked pump monitors to provide both a realtime appreciation of pump condition as well as groundwater levels across a region. Realtime monitoring of access to water and the quality of that water along with sanitation collection monitoring and optimisation for resource recovery on a regional and national basis allied with the active monitoring of spending and its impact.

In a number of cases, these approaches are already at the trial stage. For example, active sewerage network monitoring ensures a constant flow level through the day allied with a pre-programmed facility to minimise sewage blockages and monitoring for potential sewage build-ups. This may be linked to pump condition monitoring and for clearing potential blockages, along with factoring in for rainfall to maintain constant flow. This is done by using the network's capacity to hold water within the system, in order to minimise stormwater surges and discharges from CSOs (Grieveson, 2017).

Many of these elements are interlinked. For example, inland water quality monitoring and the performance of sewage treatment works and overflows. Extensive monitoring such as outlined above may, in the shorter run, be detrimental for the utility as it has to effectively comply with its environmental and service obligations, rather than seeking to sidestep them where feasible. In the longer run, comprehensive and effective compliance has the advantage of building public trust in the utility and avoiding fines and other penalties.

The ideal is a utility that is fully informed and able to act on this information, so that the best possible service is delivered with the lowest water consumed at the lowest price to the customer. Moreover, there would be effective environmental compliance and asset maintenance that does not impinge upon its customers. This ideal is some way off, but that would be the outcome if these various endeavours could be coherently integrated.

None of these objectives are impossible to attain, providing the necessary actors have a pressing reason to adopt them. They need legal and regulatory incentives to encourage utilities to go further in considering their role at the river basin or catchment area level. The same applies for industrial and agricultural users. Water and wastewater services have long faced the challenge of integrating disparate elements into a coherent whole, where each process works effectively with all the others involved. Smart water involves similar challenges, with a greater degree of complexity.

The greatest practical challenge will be a move from 'big data' to 'massive data' as various sources of information are integrated. It is already evident that the effective presentation and prioritisation of information is an essential element in relatively simple smart water systems. These challenges will be multiplied as various data sources are brought together.

10.3 The Impact of Smart Water

10.3.1 Irrigation

Smart irrigation is concerned with making extant or planned assets operate in the most effective manner possible. Smart irrigation is part of a number of ways of improving irrigation efficiency. For example Jägermeyr et al. (2015) estimate that annual global irrigation water withdrawals were 2,469 km^3 in 2004–09, of which 1,212 km^3 is returned to rivers (including through leakage from irrigation canals), 649 km^3 is beneficially consumed (transpiration) and 608 km^3 is non-beneficially consumed (evaporation). Replacing surface irrigation by drip irrigation would reduce non-beneficial consumption by 76%.

As less water is needed in a more efficient system, less needs to be abstracted as loss through return flow will be reduced. Jägermeyr et al. (2015) forecasts that all-drip irrigation would need to abstract 877 km^3 of water to deliver 605 km^3 of water for beneficial consumption with river return of 110 km^3 and non-beneficial consumption of 162 km^3. They also believe that this would on its own improve yields by 9–15%. This does not take into account water which is lost due to its inefficient application in relation to the weather and underlying soil moisture. As noted in Chapter 7, smart soil moisture monitoring systems are achieving 40% savings, which indicate the prior extent of losses. In addition, some further water may be needed for periodic flushing in order to prevent saline build-up.

It has also been outlined how smart irrigation can be a tool for reducing water consumption which also offers higher crop yields. Lower water consumption per irrigated hectare means that where resource conflicts do not exist, there is also a potential for supplemental irrigation, which has the potential to boost crop yields by 56% (Jägermeyr et al., 2016).

The figures in Table 10.1 are illustrative. Firstly, considering irrigation, the data generated by Jägermeyr et al. (2015) has been blended with the FAO estimates and forecasts. The FAO assumes 3,100 km³ per annum of irrigation water being consumed today and rising to 4,500 km³ by 2030 (FAO, 2010). These are the business-as-usual (BaU) figures. For the 2030 forecasts, the impact of moving to universal drip irrigation has been considered and then soil moisture sensing integrated with drip irrigation.

The difference highlighted in Table 10.1 for drip irrigation and drip irrigation combined with weather and soil moisture monitoring reflects the savings when ensuring water only reaches the roots when it is needed. The smart application of drip irrigation has the potential to ensure that global water abstraction can be maintained at below the 4,200 km³ pa of accessible renewable water flows.

10.3.2 Smart Water and Overall Demand

As mentioned above, another factor is eliminating hunger and respecting people's choices, for example, in wishing to have a more mixed diet. The global population was estimated at 7.4 billion in 2016, and is forecast to rise to 8.5 billion in 2030 and to 9.7 billion by 2050 (UN DESA, 2015).

Three scenarios are presented in Table 10.2. 'Dumb' is business-as-usual, which is based on the FAO forecasts. 'Smart' considers the impact of smart water management and 'augmented' assumes the beneficial application of 30% more irrigation water due to its availability through other savings.

Here, municipal network losses are forecast to be reduced from 30% to 20% and smart metering to reduce demand overall by 12% after factoring in the proportion of households with traditional meters as well as the scope for people to modify their water consumption. Companies, especially those who are operating plants in international locations, are facing increasing pressure to minimise their water footprints. Industrial water demand is assumed to be reduced by 10% as a result of this, as well as the continuing need to improve water use efficiency overall.

Table 10.1 The potential impact of smart approaches on irrigation water consumption.

Km³ pa	Current – BaU	2030 – BaU	2030 – Drip	2030 – Smart
Return	1,525	2,225	625	150
Non-beneficial	775	1,125	325	250
Beneficial	800	1,150	1,150	1,150
Total	3,100	4,500	2,100	1,550

Sources: Data developed by the author using FAO (2010) and Jägermeyr et al. (2015 and 2016) and best practice examples for smart irrigation.

Table 10.2 The potential impact of smart water on overall water consumption.

Km³ pa	Current – Dumb	2030 – Dumb	2030 – Smart	2030 – Augmented	2050 – Augmented
Municipal	600	900	730	730	830
Industrial	800	1,050	945	945	1,075
Irrigation	3,100	4,500	1,550	2,015	2,295
Total	4,500	6,450	2,865	3,690	4,200

Source: Data developed by the author from Table 10.1, FAO (2010) and best practice identified for smart approaches.

This suggests that the potential to avoid using the amount of water currently being consumed until at least 2050 is feasible. Water saved from inefficient irrigation can be kept within the 'natural' water cycle (to restore inland water flows), re-allocated to municipal or industrial use or released for new irrigation projects. This depends on the locality.

Smart irrigation cannot make exercises such as wheat growing in Saudi Arabia justifiable, let alone sustainable. It will have a greater impact on more marginal areas. For example, it would be interesting to see how smart-managed drip irrigation for cotton crops in central Asia would affect the fate of the Aral 'Sea'.

10.3.3   Smart Water and Spending

GWI (2016) reviewed the scope for 'digital water' to lower operating and capital expenditure in 2016–20, identifying potential savings of 11% overall for capital and operating spending. In both water and wastewater, greater savings were identified for treatment rather than for distribution or collection. While areas such as leakage management have had a higher profile, it may well be that the effective use of demand management will have a greater impact in lowering the amount of new assets needed.

It is likely that greater savings will be made as more smart approaches migrate into utility systems and they become increasingly integrated. Capital and operating spending savings will overlap where innovations on one side drive down costs for the other. Savings for agriculture are difficult to quantify when water is not charged for and pumping costs are subsidised. Table 10.3 considers the potential for savings in three ten-year periods against estimates of capital and operating costs in 2016.

These figures are illustrative and will vary greatly from utility to utility depending on their current circumstances and future development. They also factor in the potential for demand management to minimise the development of assets to where they are needed under lower consumption regimens. Different activities also affect each other, so that increased water reuse drives down water abstraction costs.

Conclusions

There are two overarching challenges for water management and its availability. Firstly, the availability of water and secondly the availability of the funds needed to address the current and future deficits. Smart water has an important role in addressing these

Table 10.3 Potential combined capital and operating spending savings from smart water approaches compared with current costs.

Potential savings	2016–2025	2026–2035	2036–2045
Water abstraction and treatment	15%	20%	25%
Water distribution	12%	18%	25%
Wastewater collection and drainage	10%	15%	20%
Wastewater treatment	15%	25%	35%
Wastewater recovery	15%	30%	50%
Total	13%	19%	26%

Source: Author's projections.

challenges, not as a 'cure-all' but as a part of a more coherent and sustainable approach to water management. So, the most dramatic savings in irrigation arise from the adoption of drip irrigation. Smart water allows the benefits of drip irrigation to be fully realised.

Smart metering remains the public face of smart water and will continue to do so for the foreseeable future. For the most sophisticated customers, a comprehensive suite of water use information will be an attraction. How it is in fact used will be another thing. What matters is that this information is available, and that customers are able to use as much as they feel is of actual use to them. For the utility, it will become a tool for informing and interacting with customers, based on a full and timely understanding of how its assets are operating and the immediate and longer-term challenges faced.

References

Anwar (2002) NTT DoCoMo and m-commerce: A case study in market expansion and global strategy. Thunderbird International Business Review 44 (1): 139–164.

Cox J (2017) Holding Thames Water to account. Utility Week, 23rd June 2017.

FAO (2010) Towards 2030/2050. UN FAO, Rome, Italy.

Gartner (2014) Gartner Says Annual Smartphone Sales Surpassed Sales of Feature Phones for the First Time in 2013. Press release, 13th February 2014, Gartner, Egham, UK.

Grievson O (2017) Smart Wastewater Networks, from Micro to Macro. Water Innovations, July 2017, 24–26.

GWI (2016) Chart of the month: Digital water savings for utilities. Global Water Intelligence 17 (12): 5.

Jägermeyr J, Gerten, D, Schaphoff S, Heinke J, Lucht W and Rockström J (2016) Integrated crop water management might sustainably halve the global food gap. Environmental Research Letters 11: 7–14.

Jägermeyr J, Gerten D, Heinke J, Schaphoff S and Kummu M and Lucht W (2015) Water savings potentials of irrigation systems: global simulation of processes and linkages. Hydrology and Earth System Sciences (19): 3073–3091.

Karath K (2016) World's first city to power its water needs with sewage energy. New Scientist, Daily News, 1st December 2016.

Nokia (1996) First GSM-based communicator product hits the market. Nokia Starts Sales of the Nokia 9000 Communicator. Press release, 15th August, 1996, Nokia Oy, Espoo, Finland.

Ofwat (2017) Delivering Water 2020: Consulting on our methodology for the 2019 price review. Ofwat, Birmingham, UK.

Prescott M (2017) Water Industry Risks Briefing 2017. Environmental Rating Agency, Oxford, UK.

Sedlack D (2016) The Limits of the Water Technology Revolution. Discussion Paper, Megatrends Workshop, National University of Singapore, February 24th, 2016.

Thames Water (2017) Annual report and financial statements 2016/17, Thames Water, Reading, United Kingdom.

UN DESA (2015) World Population Prospects: The 2015 Revision, Key Findings and Advance Tables. Working Paper ESA/P/WP.241. United Nations, Department of Economic and Social Affairs, Population Division, New York.

Conclusions

The business of smart water and wastewater management has evolved dramatically over the past decade. It will continue to so for at least the next decade as more early stage offerings are adopted, allowing a better appreciation of its current and potential impact.

For now, smart water exists as a series of approaches, each to some extent operating in isolation. These range from customer meters and smart showers to flood vulnerability modelling and soil moisture monitoring. Some of the greatest benefits smart water can offer will be realised when such apparently disparate tools can be connected into coherent entities at the river basin and indeed the national level.

It is evident that much remains a work in progress. Likewise, the regulatory and policy support that markets such as smart energy and communications currently enjoy are appreciably weaker for smart water. This may be a concern if, for example, smart water is treated as an afterthought to smart city projects, rather than one of the central components, and much remains to be done to ensure the interoperability of various smart water systems. Practitioners in water and wastewater have one advantage here; they are well used to being overlooked when it comes to perceived utility and infrastructure priorities.

Compared with many innovations in water and wastewater management, smart water offers the potential to be a genuinely disruptive development through the effective integration of a number of incremental improvements. Previous disruptive innovations such as slow sand filtration, activated sludge, reverse osmosis and the membrane bioreactor have been based on the continual development and refinement of a single innovation over a considerable period of time. While individual elements of smart water such as network pressure management and smart irrigation may be seen as disruptive, real disruption will take place as a number of incremental improvements in service delivery start to beneficially complement each other.

While many smart water processes remain at an early stage of development and are unlikely to be broadly commercially adopted before 2025–35, some are already being used by a significant number of utilities. What will be of particular interest will be to see if the traditional 15–25 year time lag between invention and adoption in the water sector will be changed by smart devices. One encouraging element here is that when market surveys are updated, the tendency has been to increase both the current and forecast market size estimates.

The caution and conservatism that characterises many aspects of water and wastewater management is in part driven by the service, public health and environmental

Smart Water Technologies and Techniques: Data Capture and Analysis for Sustainable Water Management, First Edition. David A. Lloyd Owen.
© 2018 John Wiley & Sons Ltd. Published 2018 by John Wiley & Sons Ltd.

obligations that these services entail. This may be challenged by the capital and operating spending savings identified for various smart approaches as well as their potential to improve operating efficiency and service delivery.

The more water tariffs (and water and energy fees for agricultural water abstraction and pumping) reflect the actual costs of providing and managing this resource, the greater the motivation for consumers to consider their consumption. Fair and equitable water pricing may in time become of the principal drivers to the broad adoption of smart water approaches.

Perhaps the only elements we can forecast with certainty is that smart water will become an integral part of water management, and that while many of the eventual outcomes will not be those that are being predicted today, they are likely to drive costs down while improving service delivery.

Index

Smart Water Technologies and Techniques: Data Capture and Analysis for Sustainable Water Management,
First Edition. David A. Lloyd Owen.
© 2018 John Wiley & Sons Ltd. Published 2018 by John Wiley & Sons Ltd.